ULTRASONIC VELOCITY STUDIES IN LIQUIDS AND THEIR CO-RELATION WITH THE STRUCTURAL ASPECTS

Ultrasonic Velocity Studies in Liquids and Their Co-relation with the Structural Aspects

Dr. SINDHU SADHU

Gian Publishing House

29/6 SHAKTI NAGAR • DELHI-110007 • INDIA

GIAN PUBLISHING HOUSE
29/6, Shakti Nagar, Delhi-110007

ISBN 81-212-0075-X

PRINTED IN INDIA

Published by Mrs. Gayatri Garg for Gian Publishing House, Delhi and Printed at Mohan Printing Corpn., Padam Nagar, Delhi.

Foreword

THE AUTHOR has conducted very useful, work in the field of molecular acoustics. Co-relation established with structural aspects of certain organic liquids is of far reaching significance.

Interesting feature of the work is that these investigations have been made with a very simple apparatus which can be easily manipulated even by a postgraduate student.

Every careful analysis of the results reported so far in the literature and those obtained by the author (some of which has been obtained for the first time) deserve special mention.

I am confident that the work as envisaged in the book will be very useful to the research workers and the senior students in the field of molecular acoustics.

Professor & Head,
Department of Chemistry,
Govt. Girls' College
Rewa (M.P.)

Dr. N.L. JAIN
M. Sc.
Ph.D. (Glasgow)

Preface

IN THIS book an attempt has been made to give an uptodate review of the work carried out so far in the field of molecular acoustics pertaining to the measurement of velocity, absorption coefficient, temperature coefficient of velocity and other constants in several mixture of chloroparaffines with electron donor liquids. Correlation has been established with the structural aspects of these liquids. A very careful analysis has been made of the results reported by various workers in the field and those obtained during the course of this work and some of these studies have been made for the first time.

The apparatus used is rather simple and elegant providing an accuracy of the order of .2 mt/sc velocity measurement. A detailed description of the manipulation of this apparatus has been given so as to make the book really useful for research workers in the field and also the postgraduate students who want to conduct some simple experiments on this apparatus pertaining to velocity measurement, compressibility studies etc.

It is earnestly hoped that the book would prove useful in the field of molecular acoustics.

C 1/79, Ashok Vihar II,
Delhi-110052
November 1986

SINDHU SADHU

Contents

Contents

1

Introduction

VELOCITY DETERMINATION in the case of ultrasonic waves of small amplitude has acquired importance because it enables us to study the properties of the medium.

Aim of the work done so far in this direction has been the following:

(1) To explain the propogation of sound waves in liquids, different theories have been suggested and expressions derived.

(2) To study the temperature coefficient in various liquids and work out the important characteristics of a substance, like compressibility, specific heat, molecular radius

(3) To establish correlation between sound velocity and other important physiochemical constants such as Molar refraction, heat of evaporation and surface tension.

(4) To study the structure of solutions and also the physical and chemical processes involved.

(5) To observe the change of phase by measuring ultrasonic velocity.

(6) To study interaction in binary and tertiary liquid mixtures.

Molecules in a liquid can be treated like rigid balls-sound wave travelling through the intermolecular space. Molecules thus short circuit all but a fraction Lt/L of the path (L-Intermolecular separation: Lt-free path). Sound velocity therefore is given by,

$$C \text{ liquid} = \left(\frac{V}{V_f}\right)^{\frac{1}{3}} C \text{ gas} = \left(\frac{V}{V_f}\right)^{\frac{1}{3}} \left(\frac{VRT}{M}\right)^{\frac{1}{2}} \quad ...(1)$$

Where V_f is the free volume

V the molecular volume

& V, R, T, M have their usual significance

Kittle[1] applying Tonks[2] equation of state to the sound propogation in liquids derived the relation,

$$C \text{ liquid} = \left(\frac{V}{V_c}\right)\left(\frac{3 \text{ r liq}}{\text{r gas}}\right)^{\frac{1}{2}} C \text{ gas}$$

$$= \left(\frac{V}{V_a}\right)^{2} \left(\frac{3RT}{M}\right)^{r} \text{liq} \quad ...(2)$$

Va being the available molar volume defined as

$$Va = V_1 + V_2 + V_3$$

V_1 = Molar volume change on heating solid from 0° k to melting point.

V_2 = Molar volume change on fusion

V_3 = Molar volume change on heating the liquid from melting point to the temperature concerned.

Va was calculated by using experimental values of sound velocity and was found to agree with that calculated from thermal expansion data.

Jacobson[3] using the concept of available volume obtained an empirical relation between sound velocity and free length, if,

$$L_f = \frac{2\,V_a}{(36\pi N V_o^{2})^{\frac{1}{3}}} \qquad ...(3)$$

Vo being the volume of spherical molecules at absolute Zero. His relation reads as :

$$CL_f = K \qquad ...(4)$$

K being a temperature dependent constant which was further elaborated by Erymig[4] in the following expression :

$$K = \left(\frac{6}{N^2\pi}\right)^{\frac{1}{3}} \left(\underset{gas}{r}\ RT\right)^{\frac{1}{2}} \nu^{-\frac{1}{6}} \qquad ...(5)$$

The quantity $\left(\underset{gas}{r^{\frac{1}{2}}}\ V^{-\frac{1}{6}}\right)$ differs only slightly from liquid to liquid. The numerical value of K as calculated by Wade[5] was, however, twice the experimental value, discrepancy being attributed to the use of Eyring liquid model, According to Lattice theory[6–7], the number of holes *NH* present in the total number of molecules N is :

$$N_H = N_e^{-v/KT} \qquad ...(6)$$

K representing the Boltzman constant, U representing activation energy i.e. the minimum work needed for removing a molecule from its lattice position.

On the basis of this model, velocity can be expressed as :

$$C^2 = \frac{V_o\, r\ KTe^{U/KT}}{\rho\ N\,(\triangle v)^2} \qquad ...(7)$$

indicating size of a hole.

The same model has been found by Hall[8] to be quite successful in explaining the absorption in associated liquids.

Schaaff[9] obtained the following relation on the basis of Van der waal's equation :

$$b = \frac{M}{\rho}\left[1 - \frac{RT}{Mc^2}\left(\sqrt{1 + \frac{Mc^2}{3RT}} - 1\right)\right] \qquad ...(8)$$

Constant b which is connected with volume has been found to

be additive. Further, an agreement has been found between the experimental and theoretical values in the case of this constant.

Arakawa[10] et al investigated the concept of free volume by the correlation function method.

With the development of precise methods (Diffraction, interferometric and pulse technique), for measurement of velocity a number of organic liquids were investigated by various workers.[11–16] These investigations established that ultrasonic velocity in liquids depends on (1) Type of liquid (2) Its density (3) Length of the Molecule (4) Electric moment of the molecule (5) Number of Benzene rings in case of Aromatic compounds (6) Viscosity (7) Presence of heavy atoms (8) Presence of double bonds. Out of these increase in length, electric moment, number of benzene rings and viscosity increase the velocity, whereas increase in density, introduction of heavy atom and double bond (unsaturation) tend to decrease the velocity in liquids.

Legemann[17] et al showed that the effect of substitution of flourine, for hydrogen in twenty organic liquids results in lowering the ultrasonic velocity.

Parthasastry has shown in general (apart from a few exceptions in which heavy atoms are substituted) that lengthening of the molecular chain enhances the sound velocity. A study of esters by him showed a contrary behaviour of decrease of velocity with increase in the molecular weight. Legemann[18] showed that in case of alkylaryl ketones, an increase in molecular weight causes decrease in sound velocity within the limits of this series. When increase in molecular weight is accompanied by Chain branching, a considerably greater drop in velocity is noted. For the Ketone series Legemann[19] observed a decrease in velocity with increase in molecular weight. Marks[20] made velocity studies in different types of alcohols. In primary alcohols, he observed a monotonic increase in velocity with increase in the number of carbon atoms,

$$\frac{\Delta c}{\Delta x} = 6.306e^{-0.15x} \qquad ...(9)$$

$\triangle_c$ being change in velocity and $\triangle_x$ the increase in the number of carbon atoms.

Ram Prashad[21] on the basis of the theory of inter-molecular attractive forces, explained Parthasastry's empirical rules between sound velocity and constitution of pure organic liquids.

According to him, it is the compressibility which determines the structure of a particular liquid.

Isothermal or adiabatic compressibility is given by $C^2 = \frac{1}{\rho\beta}$, ρ being the density of the medium and β the compressibility.

In case of molecules, according to him the following forces are present :

(*a*) General or Dispersion forces or London forces.

(*b*) Orientation or Keesom forces.

(*c*) Induction or Debye's forces.

It was shown that there can be attraction between a polar and non-polar molecule by virtue of the non-polar molecule being induced upon by the polar one prior to attraction. These forces are independent of temperature. In addition to these attractive forces, there are repulsive forces also which are significant at very small distances. Thus, internal energy, E between two molecules is given by :

$$E = \frac{C_r}{R_n} - \frac{C_a}{R^6} \quad \ldots(10)$$

n being an integer of the order of 9 to 12, R the inter-molecular distance, Cr & Ca the constants for attractive and repulsive fields respectively. Ram Prashad has further observed that the compressibility β can be assumed to be inversely proportional to internal energy E. With these observations these rules were successfully explained by him and new rules were introduced between diamagnetic susceptibility and velocity.

Schaffs[22] related the sound velocity with the packing of matter which is a microscopic arrangement of ions, atoms or molecules in a lump of matter with the aid of Van der waal's equation of state :

$$C\ \text{liquid} = \frac{3}{1\text{-}r}\left(\frac{3}{v\ \text{gas}}\right)^{\frac{1}{2}} C\ \text{gas} \qquad \ldots(11)$$

He got the following new functional relationship:

$$C\ \text{liquid} = C\left(v,\ M,\ P, r\right) \qquad \ldots(12)$$

Where r stands for packing factor and obeys the relationship:

$$h = 4r\,\frac{M}{P} \qquad \ldots(13)$$

This has provided a new method for calculating the accurate value of the packing factor r and the molecular radius given by :

$$R_m = \frac{3}{4\pi}\left(\frac{rv}{N}\right)^{\frac{1}{3}} \qquad \ldots(14)$$

These results agreed with those obtained by optical refraction method, where :

$$r = \frac{n^2 - 1}{n^2 + 2} \qquad \ldots(15)$$

and also the thermodynamic method at the critical point of the liquid, where :

$$4\,rv = \frac{RT_c}{8\,\rho C} \qquad \ldots(16)$$

As for example, Molecular radius of Benzene at 20°C by these three methods is found to be 2.01×10^{-8} cm., 2.16×10^{-8} cm., 2.18^{-8} cm. respectively. The packing factor r can also be obtained by addition as :

$$\tfrac{4}{3}\pi R^3{}_m\,N = B = \Sigma_2\,(ZA)_2 \qquad \ldots(17)$$

Where Z = Number of atoms of the same kind and

A = Atomic volume.

This suggests a straight line plot of $C\,\frac{M}{\rho}$ versus B,

Schaff tried it for 92 different aliphatic compounds containing carbon, hydrogen, oxygen, nitrogen and chlorine atoms and straight line plot was confirmed.

For each homologous series, a simple relationship existed between c liq. and r of the form :

$$C \text{ liq} = W_r \qquad ...(18)$$

W being a constant for each series depending upon efficiency of momentum transfer by collision between molecules. The molecular volume B and sound velocity could, thus, be obtained experimentally and theoretically using the relations

$$B = \frac{1}{W} C \frac{M}{\rho} \ \& \ B = \Sigma_i (zA)_i \qquad ...(19)$$

$$C = \frac{W}{M/\rho} \left[\Sigma_i (zA)_i \right]$$

The two sets of values for B & C were found to be in good agreement. Further consideration of space filling factor ruled Schaffs to assure that relation b = 4B, 4 was not the correct factor, instead he proposed that b = SB.

Where S is the collision factor. Combining r = B/v and b = SB and on substituting in the equation :

$$C = \left[\frac{VRT}{M} \left\{ \frac{G}{(1-{}^b/_v)^2} - \frac{2}{(1-{}^b/_v)} \right\} \right]^{\frac{1}{2}} \qquad ...(20)$$

he obtained the relation :

$$C = \left[\frac{VRT}{M} \left\{ \frac{G}{(1\text{-}S_r)^2} - \frac{Z}{(1\text{-}S_r)} \right\} \right]^{\frac{1}{2}} \qquad ...(21)$$

By plotting C as a function of Sr, it can be found that as M approaches infinity in a homologous series, velocity tends to the values $U\alpha$ = 1600 mt/sec. Nulsch-Kuhnldes[23] extended the Schaffs relation namely C = $U\alpha$ Sr = 1600 Sr to mixtures. He tried about ten mixtures and found that experimental and theoretical values agreed with each other.

Joern[24] also utilised the Schaffs relation for mixtures. The relation between sound velocity and $\frac{\eta^2 - 1}{\eta^2 + 2}$ was observed for mixtures of water — 2 — chloroethenol.

Kurt Attenberg[25] gave a theoretical basis of the emperical equations of Schaffs, namely $CV = K \Sigma_i (ZA)_i$ and $B = \Sigma_i (ZA)_i$.

from the consideration of inter-molecular forces and he derived the relation.

$$C = 5.663 \frac{\rho^2 P^2}{M^2} \left(\frac{N}{\rho^2 M}\right)^{\frac{1}{6}}$$

$$C \left({}^{Cm}/\text{sec} \right) = 5.663 \left(\surd \right)^{\frac{1}{6}} \left(\frac{N}{\rho^2 M}\right)^{\frac{1}{6}} \qquad ...(22)$$

When N = Avagadro number r = surface tension. Expressing this relation in terms of the Parachor P, the above relation changes to,

$$C = 5.663 \frac{\rho^2 P}{M^2} \left(\frac{N}{\rho^2 M}\right)^{\frac{1}{6}} \qquad ...(23)$$

This relation has been found to give better results for temperatures close to the solidification point.

While determining the ratio of specific heat r by velocity measurement, certain quantitative facts relating the sound velocity to structure became well known. These showed that velocity of sound itself was not an additive and constitutive property and so no applications of measurement of velocity of sound to elucidation of molecular structures seemed possible until Rao[26] discovered a function independent of temperature. By considering the linear change of velocity of sound with increasing temperature α_c and linear change of volume α_v of the liquid with temperature, he established the simple empirical formula $\frac{\alpha_c}{\alpha_v} = -3$ which can be reduced to the equation,

$$C^{\frac{1}{3}} \frac{M}{\rho} = R$$

Where R is a constant which has been found to be independent of the temperature of normal liquids and is found to possess additive property with respect to the composition and structure of molecules. R has been called molar velocity of sound or Rao's constant in analogy with molar refraction.

Number of workers[26] have provided theoretical explanation to Rao's formula on the basis of the phase rule and the Kinetic theory of liquids. Rao' constant was found to be expressible as a function of molecular weight M by the equation :

$$R = \alpha M + \beta \qquad \ldots(24)$$

Parthasastry[26(h)] modified the above equation to get a new expression between sound velocity and viscosity viz.,

$$C^{\frac{1}{3}}\rho = A + B/n^{\frac{1}{2}} \qquad \ldots(25)$$

According to this relation plot of $1/n^{\frac{1}{2}}$ versus $C^{\frac{1}{3}}/\rho$ shall be a set of straight lines for various series all emerging from the same point. This has been found to be true for the series of aliphatic hydrocarbons, fatty acids, aliphatic alcohols, esters, alkyliodides etc.

Legemann and Dunbar[27] showed that a relation of the form R = A + Bx existed between Rao's constant and other physical properties like molecular refraction, Parachor, Sounders viscosity constant, Vanderwall's constant b. Molecular magnetic rotation T, and critical volumes V_c R versus these physical constants were plotted for paraffins, esters of acetic acid, Monohydric alcohols and benzene hydrocarbons. These gave straight line plots, with nearly identical slopes. Log R plotted against boiling point was also found to give a straight line graph for all the series excepting monohydric alcohols. A and B were computed by the method of least squares.

Weisler[28,29] made a comparative study of Rao's constant and velocity of light (i.e. refractive index) for alcohols, cyclic compounds etc., and observed that Rao's constant is an additive property.

Schaffs introduced another additive quantity $\left(\frac{C}{C\alpha}\right)^{\frac{1}{3}}\frac{M}{P}$ called sound volume, C α being a constant. A comparative study of sound volume with Rao's constant and of Rao's constant with respect to the bond increment values were also made. A number of workers[30–33] calculated bond increment values for Rao's constant and found good agreement between the constant obtained by addition of such values and that from experimental data on velocity and density. Rao's expression was extended to mixtures and solutions also. In many cases where molecules of the components are not heavy, Rao's constant as a function of molecular fraction has been found to show a good linearity. In solutions also, this was observed.

Recently, Salakhuldenov[34] has shown that in place of Rao's constant $R = C^{\frac{1}{3}} V$, the additive function for velocity in liquids is more accurately given by,

$$L = R\,(M/T_c)^{\frac{1}{6}}$$

This was extended to mixtures also such as hydrocarbon—hydrocarbon and hydrocarbon-alcohol and velocity calculated was found more close to experimental values than that calculated from Rao's constant.

Rama Rao[35] using the additive and temperature independent nature of Rao's constant. Similar to molecular critical volume, derived the following expression by setting up the dimensional equation:

$$R^1 = R\,(M/T_c)^{\frac{1}{6}}\,V_c$$

The constancy of R^1 has been found to be very much better than that of R/VC.

Weisler[36] found that Molar sound velocity should be constant for associated liquids, if instead of the formula weight, we use the effective molecular weight. Thus, for a given liquid $\rho/C^{\frac{1}{3}}$ should be a relative measure of association at various temperatures. He used 0°C as a reference point for water and then defined the relative association R_A as proportional to $\left(\frac{Co}{c}\right)^{\frac{1}{3}}\frac{\rho}{\rho o}$ where subscript refer to values at 0°C. As a base, R_a for water is taken as unity at 0°C and at 20°C it is taken to be 0.98.

The determination of the velocity of sound helps in finding out many other parameters also e.g. compressibility, heat capacity and its ratio, Internal pressure[37], surface tension[38], Van der waal's constant[39], b Parachor[40], constant a of the Van der waals equation of state[41], surface energy[42], dipole moment[43] and free energy etc. All these are connected with the structure of molecules in liquids. For instance, Weisler[44] has found that for chlorides of elements of group IV the compressibility versus molecular weight curve and freezing point versus molecular weight curve are exactly identical. This indicates that compressibility has a relation with freezing points.

Rao[45] suggested that velocity of sound and compressibility follow the law corresponding states a plot of ratio of velocity of sound at a given temperature to that at critical temperature i.e. C/C_{cr} versus temperature gives a straight line. In other words at corresponding temperatures the following ratios are correct :

$$\frac{C_1}{C_{cr_1}} = \frac{C_2}{C_{cr_2}} = \frac{C_n}{C_{cr_n}} \qquad \ldots(26)$$

Where C_1 C_2 C_n are velocities of sound in different liquids at identical corresponding temperatures and $\frac{C}{cr_1}$, $\frac{C}{cr_2}$ refer to the same at critical temperature.

This fact has helped for providing, theoretical basis for Rao's constant and Legemann's rule according to which,

$$\frac{\Delta c}{\Delta T} \; M^{1/2} = 39.0 \qquad \ldots(27)$$

This has been shown to be a special case of,

$$M^{\frac{1}{2}} \frac{dc}{dt} = \frac{\text{Const}}{\sqrt{T_{cr}}} \qquad \ldots(28)$$

The other rule which has been verified is,

$$= \left(A \;\; V\right)^{\frac{1}{6}} \left(\frac{RT_{cr}}{M}\right)^{\frac{1}{6}} V_{cr} \qquad \ldots(29)$$

This relation is similar to that obtained by Kurt Attenberg[46] for sound velocity and Parachor which has also been derived utilising the fact that velocity follows the law of corresponding states.

Temperature variation of velocity has been studied for most of the pure liquids but *much* work has not been done in the case of mixtures. Factors governing variation of velocity with temperature for a liquid are (*a*) Distance between the molecule in a liquid (*b*) Its associations. These have opposite influences over dc/dt, temperature coefficient of velocity. For organic liquids with no appreciable association, velocity versus temperature curve is linear with a negative gradient. Water and heavy water are found to have a positive temperature coefficient upto 74°C and the curve is parabolic because of high degree of association. On the basis of a large number of experimental studies on liquids, it was possible to establish an empirical formula, $C = Co\ (1-\alpha \triangle t)$ where Co refers to velocity at 0°C and $\alpha = \frac{1}{c}\frac{dc}{dt}$. This formula has been found to be true for a number of liquids over a broad range of temperatures. However, as this relation does not hold for water, Petrea[46] used this relation for studying association in aqueous solutions.

Otpush Schennihov[47] et al have shown that the following expression connecting thermal coefficient of sound velocity, $\frac{dc}{dt}$, with heat capacity Cp and thermal expansion coefficient α holds for normal liquids,

$$\frac{dc}{dt} = -\ VM_{n/2}\left(VM\right)^{\frac{1}{2}}\left[\alpha\left\{C\rho-(_{2}k/2m)\right\}\right]^{\frac{1}{2}} \quad ...(30)$$

Legemenn gave an empirical relation $\frac{\triangle c}{\triangle t}\ M^{\frac{1}{2}} = 39\ Mg^{\frac{1}{2}}$ (Sec. deg.) which was found to be true for normal liquids near solidification points and critical points. The rule has been found to hold good for mixture of normal organic liquids also.

It was pointed out by Russian Scientists that while considering the temp. coefficient of velocity of ultrasonic waves in liquids, it is frequently forgotten that the values given for this quantity in the literature characterize the change of velocity with simultaneously changing temperature and density. In other words, $\frac{dc}{dt}$, in place of $\left(\frac{\partial c}{\partial t}\right)\rho$, ρ has been considered,

These derivatives cannot only differ in magnitude but also in sign. Sound velocity when plotted versus temp. for toluene with density and pressure constant, was found to be no longer linear. Similarly, for ethyl-alcohol that curve exhibited a minima. Hedamard[48] explained this behaviour theoretically.

The ultrasonic velocity studies provide an excellent means of studying various types of change of phase, like study in vicinity of melting point of supercooled liquids and around the critical temperature of individual liquids[48] and mixtures, transition in liquid crystals from their anisotropic region to isotropic region.[49] In all these studies, the velocity versus temp. curve exhibits abnormality on either side of the transition temperature. The curve either undergoes a deviation from linearity or exhibits a minima. Liquid crystals behave quite differently depending upon their type. In some of these crystals, a minima occurs in the vicinity of the liquid-crystalline phase transition, while in others a change of slope is observed on either side of the transition temperature. Such transitions in liquid crystals are being studied extensively these days. In study of glassy state, Kulbitskaya et al observed a positive temp. dependence of velocity in some high melting glasses.

Among the studies of solution of solids, those of electrolytes have been extensive. The allied parameters with respect to which the solutions have been studied are compressibility, apparent molal compressibility, Rao's constant, hydration number n, relative association *RA*, specific accoustic impedence z.

Usually in all electrolytic solutions, it is observed that compressibility decreases and velocity increases with increase in concentration of electrolyte But in case of aqueous soln. of salts of heavy metals like lead, thorium, uranium; velocity has been found to decrease with increase in concentration, Bacham has given an empirical expression for compressibility of a solution β in terms of that of solvent β_0 and Molar concentration (c),

$$\beta = \beta_o + AC + BC^{\frac{3}{2}} \qquad \ldots(31)$$

A and B being constants obtained from the β versus C curve.

The apparent molal compressibility ϕ of a solution is related to the compressibility by the relation:

$$\phi = \frac{V\beta}{n_2} - \frac{n_1}{n_2} \overline{V}_1^{\circ} \beta_o \qquad ...(32)$$

Where V is volume of the solution, $\overline{V_1}^0$ is partial molal volume of pure solvent; β and β_0 are the compressibilities of the solution and pure solvent respectively β, β_0 and ϕ can be isothermal as well as ediabatic. Gueker[50] found that ϕ versus square root of concentration is linear up to highest possible concentration available when the concentration is expressed in volume (i.e. number of Moles of solute per litre of solution). Extrapolation of this st. line to zero concentration (i.e. infinite dilution of solute), gives a quantity ϕ^0 which is additive property of ions. ϕ^0 changes very little with pressure and temperature and is negative for all strong electrolytes. Further, it decreases while slope increases as one goes from 1-1 to 2-1 and 2-4 valent electrolytes. Gueker[51] has derived the following expression using the Debey-Huckel's theory[52] of electrolytic solution.

$$\phi = \phi^{o} + a_1 C^{\frac{1}{2}} \qquad ...(33)$$

This has been named Gueker's limiting law. Subsequently, other workers[53–56] studied ϕ in several solutions of electrolytes of different valence types and reported deviations from this law. Gueker and Haag[57] measured velocity for solutions containing dipolar ions so as to see the effects of these dipolar nature of solute molecules on the apparent molal compressibility.

Determination of apparent molal compressibility helps in the study of the hydration. Various workers[58] used this to determine the number of molecules bound to the ions of an electrolyte in aqueous solutions. The assumption is made that the ions are incompressible in comparison with the free solvent molecules. The electrostrictive effect of the ions on the water molecules in the immediate vicinity of the ions is equivalent to a large internal pressure on these solvent molecules. According to these assumptions,

$$\lim_{c \to o} \phi = -n_h V_1^\circ \beta o$$

Where n corresponds to the number of water molecules in the primary hydration sheaths of the ions. Inherent in these equations is the further assumption that the compressibility of water molecules near but not in the primary hydration sheaths is the same as that of pure water. Left hand side of above expression can be obtained by extrapolating the ϕ versus C curve. Thus n.H. has been obtained[59–60] for various electrolytes and compared with that obtained from mobility, dielectric entropy and activity measurements. The agreement among these values have been found to be not very good.

Addition of electrolytes in moderate concentration to water makes the maximum in velocity versus concentration curve shift to lower concentration. This was explained on the basis of a model involving interpenetrating solvent lattice and ionic lattice, each of which is presumed to occupy the whole volume of the system.

Shillinger[61] treating the ion core forces as those acting between rigid spheres, has shown that the distance of closest approach 'a' for an anion cation pair can be predicted from compressibility data, using the relation:

$$P\,K_T B_T = \frac{(1-y)^4}{(1+2y)^2} \qquad \text{...(35)}$$

$$\text{where} \quad y = \frac{\pi P_a{}^2}{6} \qquad \text{...(36)}$$

$$\text{or,} \quad a = \left(\frac{6y}{\pi P}\right)^{\frac{1}{2}} \qquad \text{...(37)}$$

P being the number density of particles in the fluid, K = Boltzman constant, T = absolute temperature, BT = isothermal compressibility. 'a' so calculated has been found to be a little less than that obtained from the crystal studies. Veeraih[62] utilised the above expression for computing 'a' for molten electrolytes upto 800°C. Besides aqueous solutions, solutions of electrolytes in organic solvents were also studied by different workers[63] who found results somewhat similar to those for aqueous solutions.

Studies were also made with mixed salts, that is mixtures of the salts in one solvent. These were very interesting as some of them formed complexes while in solution. Shobby[64] et al studied the formation of complexes called chelates between organic substances and electrolytes when dissolved in the same solvent. These authors, from the velocity and compressibility studies, succeeded in calculating the stability constants log K, $\triangle H$, $\triangle f$, $\triangle s$ etc. and found them to compare well with those obtained by thermodynamic method.

Other parameters studied in elucidation of single and mixed electrolytic solutions are specific acoustic impedence z and relative association *RA*. Former is a measure of the resistance offered to the passage of acoustic waves and is given by the product of density and velocity. The impedence z increases with increase of number of ions. Relative association has already been defined by equation RA= $\left(\frac{Co}{C}\right)^{\frac{1}{3}}\left(\frac{\rho}{\rho_o}\right)$ and is influenced by (*i*) breaking of solvent molecules on addition of electrolytes which results in decrease of RA and (*ii*) hydration of ions whose influence is to increase RA. RA for mixed salt has been found to be larger than that for individual salts at the same concentration which implies that hydration taking place in former is more predominant than in individual salts. Ternary system of electrolytes has also been studied recently.[65]

Such extensive study has not been done in non-electrolytic solutions, as they do not interact with solvent molecules to that extent as electrolytes do. Some work has been done in this field with respect to temp. and concentration.

The general behaviour of all the non-electrolytic solutions with the exceptions of those studied by Rao[66] and Lal[67] is as follows : (*i*) Velocity increased with increase of concentration, (*ii*) Rao's constant varied linearly with concentration, (*iii*) Extrapolated value of sound velocity and Rao's constant for 100% solute concentration have nearly the some value for each solute irrespective of solvent and are quite close to the value of velocity and Rao's constant for solute, had it existed in liquid state at that temperature. Later velocity being obtained by extrapolation of velocity versus temp. curve in molten state to the required temperature and later Rao's constant obtained by using Legemann's bond increment method.

Lal[67] observed an unusual decrease of velocity with increase of concentration in benzoic acid dissolved in alcohols Rao and Rao[68] studying the same solution, reported result contrary to Lal. Rao[69] found p-dichloroethane to behave similar to that studied by Lal when dissolved in m-xylene and nitro benzene.

Aqueous solutions of non-electrolytes have also been studied by a few workers. Hazimeshia[70–71] worked on solutions of Seccharides in water and gave a method of calculating amount of bound water for these non-electrolytes. Yutaka Miyahera[72] made a similar study on Gelatin solutions.

According to Hazimeshia[70–71] if M grams of solute is dissolved in Vo cc of solvent, then some of the solvent (say Vo' cc) will be attached to solute and some will be compressed to V^{2}' and consequently there will result a solution of volume V'cc. so that $V' = V'o + V_1' + V_2' - Vo$, V_1' being the volume of solute in solution. Amount of bound water then comes out as :

$$\frac{\triangle}{c} = \frac{\beta_1}{\beta_0 d_1} - \left\{ (V_0 - V_2\ \beta_2/\beta_0)/c \right\} \qquad ...(38)$$

where $\beta_1, \beta_0, \beta_2$ are respective adiabatic compressibilities of solute, solvent and bound water d_1 is the true density of solute in solution, given by $d_1 = \frac{M}{V^1{}_1}$

$$V_o = \frac{V_g}{V} = \frac{d-c}{do}, \quad V_2 = \frac{V_2'}{V'} \qquad ...(39)$$

d is the density of solution and c is the concentration of solute in gms/cc of solution, first terms of right hand side in expression for is due to compressibility of solute and the second term represents hydration. The above relation after slight modification was used by Yutaka to study the effect of addition of alcohols to aqueous solutions of Glucose, Maltose and Dextrin.

Nomoto[73] studied state of association of water molecules in aqueous solution and calculated mole fraction of associated and single molecules of water for solutions of alcohols, acetone etc. from velocity data.

Ultrasonic velocity study has been made of time reactions also. Vasakuwada[74] studied the time reaction of dilute Benzene solution of Vinyl acetate. Kowalska[75], et al studied the change of velocity of sound during the time of neutralization process and observed unusuality in velocity versus time curve at the time where the phase versus time curve showed a change. Kudryavtsav[76] et al studied the reacting liquid mixtures of aqueous solution of acetic anhydride. Yasunaga[77] studied chemical Kinetics by measurement of sound velocity and calculated rate constant of the reaction.

Another field where ultrasonic velocity study has been extensive is in liquid mixture. The case of liquid mixtures is important in the theory of liquids. The velocity and compressibility study of the mixtures throws appreciable light on molecular interaction between different species of molecules and also the structure of molecules as derived from the resulting intermolecular action.

It had been a topic of great debate as to which concentration would give a linear variation of velocity versus composition curve for ideal solution. It was found that whatever concentration (weight, molar or volume fraction) is employed, the velocity versus concentration curve always deviates from linearity even in some of the ideal mixtures. According to Kaulgud[78–79], such deviation from linearity are governed by non-ideality in free length. He computed free length using Jacobson's relation and found that in cases where experimental free length was greater than ideal free length. The velocity versus composition curve is concave upwards and vice-versa. Further it is shown by him that the free length and hence velocity is linear function of surface fraction defined by :

$$S_1 = \frac{W_1\, Y_1/M_1}{W_1\, Y_1/M_1 + W_2\, Y_2/M_2} \qquad \ldots(40)$$

where W is weight fraction and subscripts 1 & 2 refer to the two components of liquids.

The first attempt that was made for explanation of behaviour of velocity versus concentration and compressibility versus concentration curve was by Ram Parshad[80–81], who, using the theory of molecular force fields, explained the velocity and compressibility variation for alcohol-water[88], alcohol-hydrocarbon[89], ether-acetone[90] and ether-chloroform[91] mixtures.

Generally, it has been found that velocity versus composition and compressibility versus composition curves have opposite signs. But there have been some notable exceptions to this also. Suckmann[77](b) et al and Kaulgud[83] have explained this for cases of simple mixtures using the theory of corresponding states. Kaulgud's conclusion regarding this is : In polar—non-polar systems, a mere knowledge of expansion or contraction of volume is not sufficient to predict the nature of velocity and compressibility deviations correctly, but magnitude and sign of excess entropy has also to be taken into account. Only in cases where excess entropy is positive and large, one can expect a negative velocity. Deviations are considerably affected by the details of the molecular interaction. Like-wise, in such cases the compressibility deviation is negative irrespective of the sign of velocity deviations which depend upon the extent of dipolar association. There have been many types of studies in mixtures from various aspects, for example : (*i*) Nomoto[84–85](a) studied about forty mixtures to see how the Rao's constant varies from mixture to mixture, (ii) Thamsen[24], Nutsch-Ruhnkics[23] used the Schaffs relation connecting sound velocity to molar refraction and molecular volume for various mixtures and got better results than Nomoto, (iii) Jacobson[3] and Kaulgud[78–79] tested the utility of Jacobson's relation for mixtures, (iv) studies were also made in series of mixtures[86–87], one of the components being a member of homologous series, with a view to see the effect of length of molecules on velocity variation, (v) Grishin[88] studied the mixtures of Benzene with various alcohols.

The other aspect of study was the influence of chemical characteristics, for example the nature of velocity and compressibility versus composition curves for following types of mixtures of organic substances: hydrocarbon-hydrocarbon,[98] hydrocarbons and polar substances with acid[89–90], alcohol-alcohol[91] etc. General observation of these workers is that when two liquids are mixed, following possibilities arise :

(*i*) Rupture of association when one of the components is polar and the other is associated, or both are associated. This results in decrease of compressibility from the ideal value.

(*ii*) Formation of loose complexes between dipoles, increasing the compressibility value.

(*iii*) Volume contraction which decreases free length and compressibility but increases density.

(*iv*) Induction forces like that of benzene-acid mixture. This results in increase of compressibility.

(*v*) One of the component repelled by the other like in CCl_4-acid, which results in increase of compressibility.

(*vi*) Dipole association, where compressibility decreases and velocity increases over the ideal value.

(*vii*) Formation of compound by chemical bonds resulting in increase of compressibility over ideal one, for example CS_2-Acetone mixture.

(*viii*) The effect of formation of H_2 bonds is not very clear and also it has not been studied very extensively.

It was pointed out in one of these studies that the compressibility of mixture will be less than the ideal one for mixture in which individual molecules form complexes due to dipole association and compressibility will be greater than ideal value in mixtures where complexes are formed due to chemical bonds.

Some work has also been done in ternary mixtures. Voleisis[92] et al have studied the ternary mixture Methylene chloride - Ethyl alcohol-water.

Studies undertaken by the author

The above review of the work done on ultrasonic velocities in liquid media confirms the view that it certainly is an important physical constant. Its accurate measurement can therefore help us in studying certain more phenomena, which have not been investigated so far.

The ultrasonic interferometer capable of measuring velocities in liquid media to within ± .2 mt/sec. & ± .5 mt/sec. in the range of the room temperature to the boiling point of water, was therefore set up. Details of the apparatus, the method of taking observations etc. appear in chapter II.

Since the phenomenon of H_2 bond formation appears to be the one in which very little work has been done, mixtures in which this is expected to occur were chosen for investigation. Chapter III deals with the results of these investigations which show the deviations in velocity and compressibility and other allied parameters due to the formations of H_2 bonds.

The influence of change in temp. on the magnitudes and signs of these deviations were further investigated and the results have been discussed in chapter III. The discussion on the validity of Legemann's rule and nature of temperature coefficient of velocities versus composition curve due to formation of hydrogen-bonds also appears in the same chapter. Finally, the utility of the temp. coefficient of velocity in studying the formation and rupture of H_2 bonds in aqueous solutions has also been discussed in this chapter.

A detailed study of research work done on the change of phase form ultrasonic velocity measurements shows that no work has been done to study the transition from solid-liquid solution to liquid-liquid solution. A number of solutions with solutes having low melting points were, therefore, studied. The results and discussions on these studies are mentioned in the last chapter.

In this chapter a relation between the sound velocity and boiling point has been derived. The validity of this relation for liquids belonging to various homologous series, has then been examined.

2

Experimental Details

ULTRASONIC INTERFEROMETER

1. Working Principle

An ultrasonic Interferometer is a simple and direct device to determine the ultrasonic velocity in liquids with a high degree of accuracy.

The principle used in the measurement of velocity (V) is based on the accurate determination of the wave length in the medium. Ultrasonic waves of known frequency n are produced by a quartz crystal fixed at the bottom of the cell. These waves are reflected by a movable metallic plate kept to the quartz crystal. If the separation between these two plates is exactly a whole multiple of the sound wave langth, standing waves are formed in the medium. This accoustic resonance gives rise to an electrical reaction on the generator driving the quartz crystal and the anode current of the generator becomes maximum.

If the distance is now increased or decreased and the variation is exactly one half wave length $\lambda/2$ or multiple of it, anode current becomes maximum. From the knowledge of wave length λ the velocity (V) can be obtained by the relation.

$$V = n\lambda$$

2. Description

The ultrasonic interferometer consists of the following parts :

(A) The High Frequency Generator
(B) The Measuring Cell

(*a*) The High frequency generator is designed to excite the quartz crystal fixed at the bottom of the measuring cell at its resonant frequency to generate ultrasonic waves in the experimental liquid filled in the "Measuring Cell". A Micrometer to observe the changes in current and two controls for the purpose of sensitivity regulation and initial adjustment of the Micrometer are provided on the panel of the high frequency Generator.

(*b*) 'The Measuring Cell' is a specially designed double walled cell for maintaining the temperature of the liquid constant during the experiment. A fine micrometer screw has been provided at the top which can lower or raise the reflector plate in the liquid in the cell through a known distance. It has a quartz crystal fixed at its bottom.

For initial adjustment two knobs are provided on high frequency generator, one is marked 'Ajj' with this position of the needle on, the Ameter is adjusted and the knob marked 'Gain' is used to increase the sensitivity of the instrument for greater deflection if desired.

The meter is used to notice the number of maximum deflections while micrometer is moved up and down in liquid.

3 Measurements

The measuring cell is connected to the output terminal of the high frequency generator through a shielded cable. Cell is filled with experimental liquid before switching on the generator. The ultrasonic waves move normal from the quartz crystal till they are reflected back from the movable plate and the standing waves are formed in the liquid in between the reflector plate and the quartz crystal.

The micrometer is slowly moved till the anode current on the meter on the high frequency generator shows a maximum,

A number of maxima readings of anode current are noted and their number N counted.

The total distance (d) thus moved by the micrometer gives the value of wavelength λ with the help of following relation:

$$d = N\lambda/2$$

Once the wavelength λ is known, the velocity V in the liquid can be calculated with the help of relation given by equation:

$$V = n\lambda$$

4. Technical specifications

(*a*) *High Frequency Generator*

Mains Voltage	: 220 V. 50 C/S
Measuring Frequency	: 2 MHz
Frequency Tolerance	: 10.05%
Fuse	: 150 MA
Size	: 42x30x14 Cms.
Weight	: 4.8 Kg. approx.

(*b*) *Measuring Cell*

Maximum displacement of the Reflector	: 20 m.m.
Required quantity of liquid	: 12 cc approx.
Least count of Micrometer	: .01 m.m.
Size	: 11.0 x 11.0 x 35 Cms.
Weight	: 4.0 Kg. approx.

For increasing the accuracy of velocity measurement following measures were taken:

(*i*) The original micrometer of least count .001 was replaced by the one with a least count of 0.0001 cm.

(*ii*) Although many readings for position of reflector at an interval of ten maxima were taken, still in the wave length calculation, only the maximum possible distance traversed by the reflector was considered.

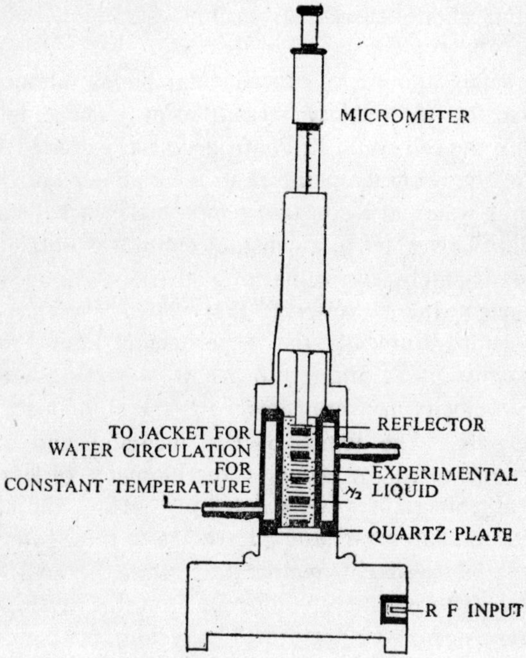

THE MEASURING CELL

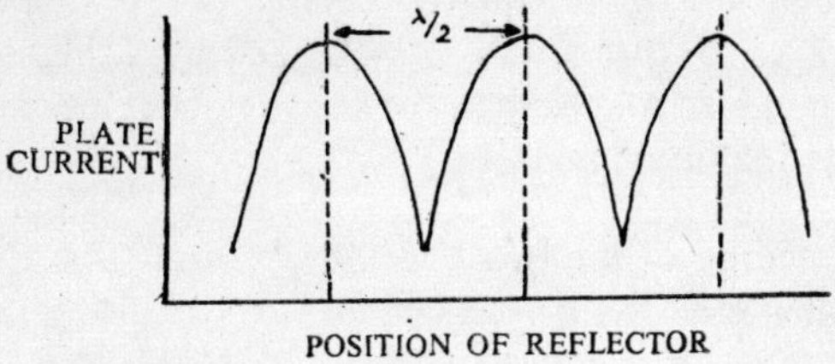

Fig 1

With these measures, it was observed that the velocity could be measured accurately to within about 0.2 mt/sec. depending upon the sharpness of the maxima and also the constancy of the temperature within the cell. It was observed that when the screw was moved by $\pm .0002$ Cm, a definite change in the value of the anode current from the maximum value was noticeable.

A very small amount of experimental liquid (about 12 c.c.) was required for the velocity measurement. The temp. of this liquid within the cell could be controlled at any desired value in the range of the room temperature to the boiling point of water, by circulating water at a constant temp. in the Jacket surrounding the cell. This water at a constant temp. was obtained from a separate assembly consisting of a thermostat and a small electric pump. It was observed that for measurements upto 50°C the temperature of the experimental liquid remained absolutely constant during the course of wave length measurement. The velocity measurements were, therefore, made mostly upto 50°C only. The temp. of the incoming and outgoing water was measured by accurate thermometer of least count 1°C. These temperatures were noted just before and after the wave length measurements and an average of these four reading was taken. As the actual temperature of the liquid inside the cell was found to be slightly different from the average value, a blank experiment was performed at various temperatures for all the velocity measurements and a correction was accordingly applied to the average value in all the cases.

The frequency of the quartz crystal, as per specification, was 2 MHZ/sec and this was also checked by using an absorption wavemeter. A specimen of the velocity measurement is given below for liquid benzene.

Initial reading of the two thermometers : 38.4°C and 38.2°C

Final reading of the two thermometers : 37.8°C and 37.7°C

Average of these four temperatures = 38°C
Correction at 38°C = 0.4°C

Hence the actual temp. of the liquid within the Cell = 37.6°C

Initial reading of the micrometer Screw : 0.0153 cm.

Final reading of the Micrometer Screw : 2.1888 cm.
for the 70th Maximum

The total distance traversed by the reflector : 2.1735 Cm. therefore, the distance between two consecutive maxima :

$$\frac{2.1735}{70} : 0.03105 \text{ cm.}$$

$$\text{Wave length} = 2 \times 0.03105$$
$$= 0.06210 \text{ cm.}$$

Sound velocity $= 0.06210 \times 2 \times 10^6$ cms/sec.

In Benzene, at 37.6°C.

= 1242.0 mt/sec.

The accuracy of $\pm$ 0.2 mt/sec. was calculated in the following manner :

With the possible uncertainty of $\pm$ 0.0002 Cm, in the initial reading, it could be .0151 Cm. or 0.0155 Cm. and similarly the final reading could be 2.1886 Cm. or 2.1890 Cm. Thus the possible maximum and minimum Value of d could be 2.1739 Cm. or 2.1731 Cm. and hence the velocity values will be 1241.77 mt/sec. and 1242.22 mt/sec. Thus, the velocity uncertainly would be about $\pm$ .22 mt/sec. This is the accuracy when the temp. of the liquid remains constant (as it happens at temperatures below 50°).

Figure 1 gives the sketch of the measuring cell and a curve expected between the plate current and the position of the reflector. Photograph of the entire assembly appears in Fig. 2.

The density at various temperature of the experimental liquids was measured by a Pyknometer of capacity 7.93 ml. The volume of the Pyknometer was previously determined by using benzene. As the boiling point data for the mixtures mentioned in chapters III & IV were not available, boiling point was also measured for all the mixtures. All the chemicals used were

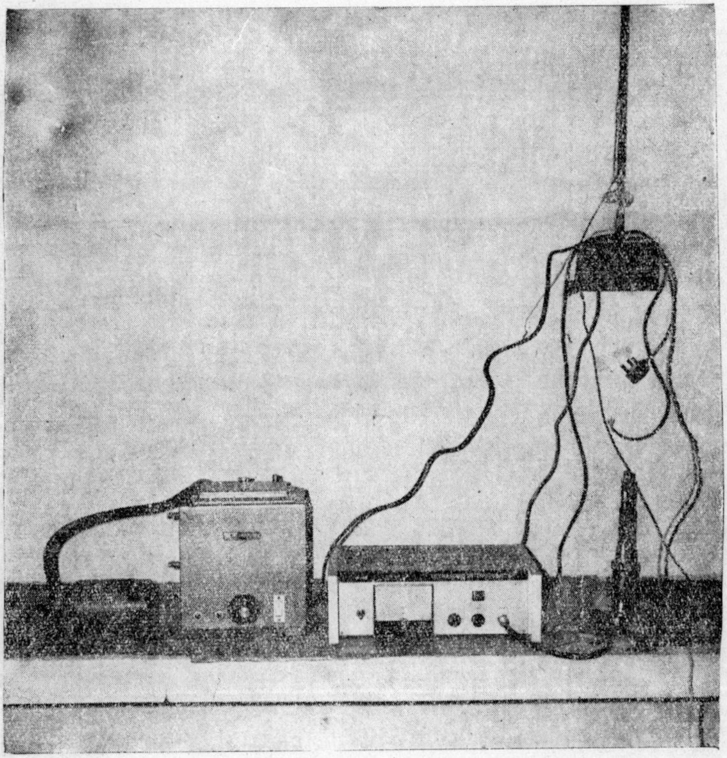

Fig. 2

of A.R. grade and freshly supplied by the various firms in India.

The concentration used for the mixtures mentioned in chapters III and IV are in weight fraction and mole fraction. As all the liquids are highly volatile, mixtures were not prepared by weighing the liquids, but were prepared by calculating the required measure of volumes of the components necessary for giving the required weight fraction, the density values being known accurately.

3

Ultrasonic Velocity and Hydrogen Bonding

PART I

HYDROGEN BOND is a bond formed between two electro-negative atoms with hydrogen atom acting as a bridge. A hydrogen atom has only 1-.s orbital, which becomes saturated after the formation of one covalent bond thus making it incapable of forming a second covalent bond. However, the positive H_2 can polarise the lone pair orbital of another atom, delocalize the lone pair of electrons and thereby form a weak long covalent bond. Thus, H-bond is a simple case of dipole-dipole (or ion-dipole) attraction. Hydrogen bonds have also been found to exist between a carbon atom and an electro-negative atom provided there are some electro-negative atoms attached to the carbon atom to activate the H-atom.

H—bonds have the following special features :

(*i*) These are very weak in nature, for example, they have an energy of only 5 kcal/mole, where as an ordinary covalent bond has an energy of about 80-100 kcal/

mole.-Strength of a H-bond has been found to decrease with an increase in the electronegativity of two bonded atoms.

(*ii*) Increase in eletronegativity of an atom increases its power of forming hydrogen bonds, *i.e.* phenols form stronger H-bonds than aliphatic alcohols because of increase in electronegativity of the oxygen atom. resulting from resonance with structures viz.  = o H+

Small size of electronegative elements is also necessary, e.g. nitrogen and chlorine have some electronegativity but chlorine is by far the poorer hydrogen bonding agent.

(*iii*) H-ion in a hydrogen bond is not centrally situated but is nearer to one of the two adjascent electronegative atoms.

Although weak in nature, hydrogen bonds are bound to effect the physical properties of the substance appreciably. For axample :

(*a*) Boiling point and melting point of substances involved in hydrogen-bonding, increase abnormally as a result of extensive molecular association e.g. water could not have been in liquid state at room temperature, had there been no association due to presence of hydrogen bonds amongst its molecules. The boiling point would have been 80°C in place of 100°C.

(*b*) Properties like heat of formation, heat of vaporization, dielectric constant, etc., also undergo abnormal behaviour when hydrogen bonds are present.

(*c*) Also, it is only due to the presence of hydrogen bond in water that the temp. coefficient of velocity in it, is positive unlike in rest of the liquids where it has a negative value.

The presence of hydrogen bonds has been detected by various means e.g. Infrared absorption spectra, X—Ray analysis, electron diffraction, boiling point and melting point study, stability examination etc.

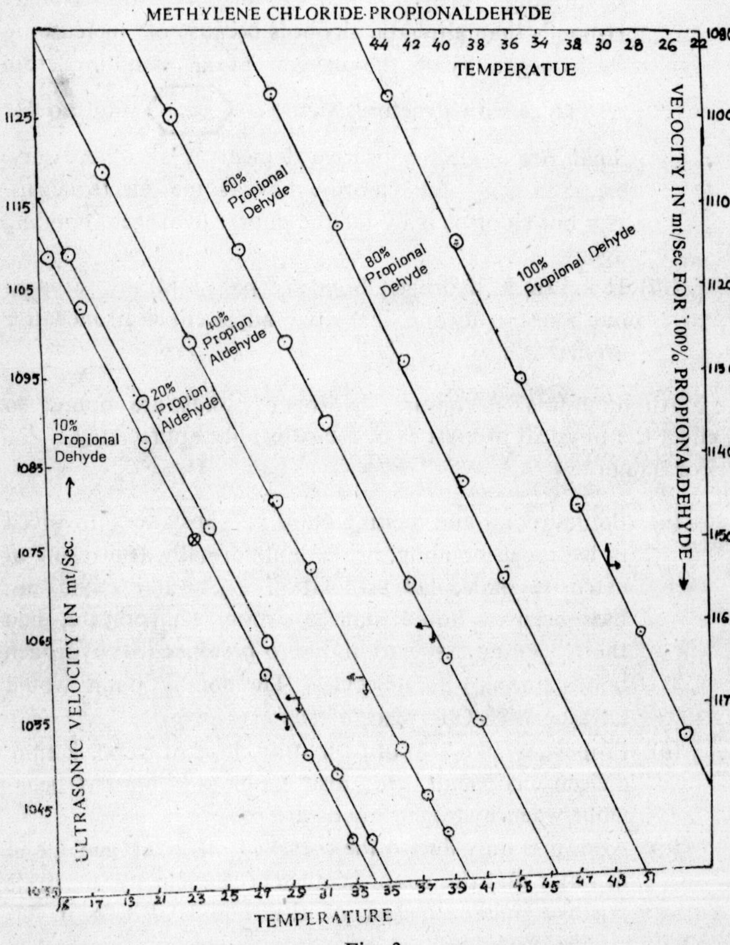
METHYLENE CHLORIDE-PROPIONALDEHYDE
TEMPERATUE
44 42 40 38 36 34 32 30 28 26 22
VELOCITY IN mt/Sec FOR 100% PROPIONALDEHYDE
1090 1100 1110 1120 1130 1140 1150 1160 1170
ULTRASONIC VELOCITY IN mt/Sec.
1125 1115 1105 1095 1085 1075 1065 1055 1045 1035
10% Propional Dehyde
20% Propion Aldehyde
40% Propion Aldehyde
60% Propional Dehyde
80% Propional Dehyde
100% Propional Dehyde
TEMPERATURE
15 17 19 21 23 25 27 29 31 33 35 37 39 41 43 45 47 49 51

Fig. 3

Very little work seems to have been done regarding influence of presence of hydrogen bonds on the ultrasonic propagation constants. A summary of this work is given below :

(1) Sunder Rao[93] and coworkers observed in acetone-water mixture an anamolous peak in ultrasonic velocity versus composition curve at about 25% by volume of acetone. The formation of compound $(CH_3)_2CO$— — —HOH through hydrogen bond was suggested as the explanation. Koteswaram found supporting evidence from Raman spectra for this mixture. Puranic[94] and coworkers studied the same mixture by refractometric methods and found a positive deviation from linearity of refractive index versus concentration curve. They attributed this behaviour to the formation of hydrogen bonds of the type $(CH_3)_2$ CO— — —HOH. The molar refraction value for O—H. . .O bond was found to be of the same order as that arrived at by Batsanov.[95]

(2) It was found that ethers and Ketones show the same type of abnormality when mixed with chloroform as far as the following properties are concerned : solubility heat of solution, viscosity, dielectric polarisation, vapour volume contraction etc. In both the mixtures, the presence of H-bonds was suggested as explanation. Ram Prashad[82] while studying those two mixtures found that :

(*i*) Ultrasonic velocity versus composition curve shows a negative deviation from linearity in both the cases.

(*ii*) However, as far as compressibility versus concentration curve is concerned, mixtures of others and acetone with chloroform behave in an oppposite manner. In accetone-chloroform mixture, the curve was found to undergo a positive deviation from linearity but for ether-chloroform mixture, a negative deviation was observed.

He explained this abnormality in compressibility versus concentration curve by pointing out that hydrogen-bonding is a special case of dipole association and in the above mixtures, both hydrogen bonding as well as dipole association occur. In the later case hydrogen and oxygen are attached under their natural field but there is no actual donation of the electrons. When two such unlike dipoles are at a distance from each other, there exists cohesive energy between them. Cohesive energy increases rapidly and hence compressibility will decrease as distance decreases. But when the distance become so small that the two molecules become a single unit through the hydrogen bond, the cohesive energy between them, thought very strong, gets ruled out from contributing to the decrease of compressibility. Now it will be cohesive energy between two such separate units that should be considered for sound propagation ; as a hydrogen bond is formed making two molecules as one close unit. The distance between the separate units will be greater than the distance that would have been there between two free molecules in the absence of hydrogen bonds. Since cohesive energy falls off as inverse of the sixth power of the distance, this will lead to increase of compressibility. Besides, the dipoles in hydrogen bond become effectively shielded for veilding any cohesive effect upon other units or molecules. In ether-chloroform mixture, dipole association predominated over hydrogen bonding, whereas in acetone-chloroform mixture, it is vice versa. Thus, Ram Prashad has shown an excellent means of finding which of the interaction (H-bond or dipole-association) predominates in a solution. Surprisingly, there has not been any further work on other mixtures to justify this statement.

(3) Fort & Moore[96] made adiabatic compressibility studies of binary liquid mixtures which involved hydrogen bonding mixtures, e.g. chloroform-diaxane, chloroform acetone, S-tetrachloro ethane-acetone, O-chlorophenoldioxane. Another set of system also involving in hydrogen bond formation but whose one component is associated in pure state were acetone-water, dimethyl sulphoxide-water, dimethyl sulphoxide-methyl alcohol, dimethyl sulphoxide-acetic acid. They observed except in the case of chloroform-dioxane, all other systems involving hydrogen bonding showed a negative excess compressibility, these being large in those cases where one component is associated in pure state.

They concluded that there is a general tendency for positive excess compressibility to decrease and become increasingly negative as the strength of the interaction between unlike molecules increases charge-transfer : dipole-dipole interactions and hydrogen bonding between unlike components should make a negative contribution to compressibility.

(4) Seshagiri[97] et al observed abnormal behaviour of velocity and compressibility versus concentration curve in mixtures of carboxylic acid with common solvents. One among these mixtures studied was chloroform-acetic mixture, where abnormality was attributed to the formation of hydrogen bonds between acetic and chloroform. These mixtures according to the classification of liquids by Ewell[98] belong to a category where H-bonds are formed as well as broken as the concentration varies. They observed that the compressibility versus concentration curve exhibits a maxima at 48% of acetic acid and minima at 80%. Velocity versus concentration curve exhibits a minima at low concentration of acid and then increases steeply corresponding velocity for pure acid. Effect of dissociation is to decrease the free length and hence the compressibility. But there is opposite effect of increase of free length produced by decreases of cohesive energy due to bond formation between chloroform and acid molecules. Hence the net effect is an increase of free length and compressibility. Initially as the acid concentration is increased, this effect increase a stage will be reached when all the acid dipoles would attach themselves to chloroform dipoles and some chloroform dipoles would be left free. At this stage, free length and hence the compressibility would begin to decrease. As acid concentration is further increased, compressibility would also decrease attaining ultimately the value corresponding to that pure acid.

(5) Persianova[99] et al, studied the mixtures of polyhydroxy alcohols with monohydroxy alcohols and with non-polar substances. According to them, compressibility in polyhydroxy alcohol is mainly due to hydrogen bonds, where as that in monohydroxy alcohol is due to the dispersion interaction of Hydrocarbon radicals of alcohols. Presence of hydrogen in

a substance has the influence to decrease the compressibility, for instance it is because of presence of such bonds that the compressibility of alcohols is somewhat lower than that of their isomeric simple ethers. While studying the mixture of glycerol, ethylene glycol with methyl alcohol, ethyl alcohol, propyl alcohol etc., they observed that addition of polyhydroxy alcohol to monohydroxy alcohol results in a negative deviation from linearity of compressibility versus concentration curve. This was attributed to formation of hydrogen bonds with molecules of the latter. This negative deviation in the region rich in monohydroxy alcohol was found to decrease while passing from lower to higher alcohol that is in the direction of decrease of degree of association and compressibility. For glycols, the compressibility deviation was found to be relatively high owing to a greater number of OH groups. The compressibility of mixtures of alcohols and monopolar substances was also found to undergo negative deviation from linear value. The observed deviation was found to decrease with decrease in the association of alcohols.

The results of these studies do not seem to be in agreement with each other. For, according to Sunder Rao et al, the effect of presence of H-bonds in a system is to make the velocity of a mixture attain a value above the linear one. Ram Prashad's observation is that the velocity versus concentration curve shows a negative deviation from linearity in such mixtures where there is either H-bonding or dipole association present, whereas compressibility versus concentration curve shows positive deviation in mixtures of former type and a negative deviation in those of later type. According to the Fort and Moore, hydrogen bonding between unlike components makes a negative contribution to compressibility where as Seshagiri et al found it to make a positive contribution to the same. On the other hand, Persianove's observation indicated that the presence of H-bond in a substance causes the compressibility to decrease. From these studies, it is not clear as to what would be the definite nature of the velocity (and allied parameters) versus concentration curve in presence of hydrogen bonds between two components of the system. In many of the mixtures mentioned above, one of both of the components were associated in pure state so that there was possibility of breaking of H-bonds as well. Hence the observed velocity and compressibility behaviour could have been the result of breaking instead of formation H-bonds. In the mixtures studied by Prashad, there is only formation of H-bonds as the components are themselves free of it in pure state and also the explanation given by him for the behaviour of velocity and compressibility in these mixtures appears very convincing. As there has not been any further work to corroborate to his views, work was undertaken by author to study velocity and allied parameters in such mixtures where there is only formation of hydrogen bonds between the components (which are themselves free of association while in pure states). From Ewells[98] classification of chemical compounds, one finds that the mixtures of chloroparaffins with electron donor liquids like ketones, aldehydes, esters, oxides etc. are of the type of above description. i.e. they themselves being free of hydrogen bonds in pure state. From such bonds when mixed, the hydrogen bonds so formed are of the type CH— — —X, where X is an electronegative atom of the donor liquid.

Ewell observed that in such mixtures the boiling point versus concentration curve goes through a maxima showing the presence of H-bonds forming power which depends, in each chloroparaffin series, upon the number of chlorine atoms in the chloroparaffin. This increases e.g. in chloromethane in the following order :

Methylene Chloride $<$ Chloroform

and in Chloro-ethers, the order is,

1,2 Dichloro ethane $<$ 1,1,2, Trichloroethane ∠

1,1,2,2 tetrachloroether $<$ Pentachloroethane...

In the present study mixtures of chloroparaffins with electron donors having oxygen atoms were chosen, the bond being CH———O. These are listed below :

	Electron Acceptors	*Electron Donors*
(*a*)	**Chloro Methanes**	
1.	Methylene Chloride	Propionaldehyde
2.	Chloroform	Acetone
3.	Chloroform	Propionaldehyde
4.	Chloroform	Ethylmethyl Ketone
(*b*)	**Chloroethanes**	
5.	1, 2 Dichloroethane	Ethyl Methyl Ketone
6.	1, 1 ,2 trichloroethane	Dioxane
7.	1, 1, 2 trichloroethane	Ethyl butyrate
8.	1, 1, 2, 2 tetrachloroethane	Cyclohexanone
9.	1, 1, 2, 2 tetrachloroethane	Mesityl oxide
10.	Pentachloroethane	Cyclohexanone

With the exception of chloroform-acetone mixture, all others were studied by the author for the first time from the point of view of examining the influence of H-bonding on ultrasonic velocity and other allied parameters. Velocity, density and boiling points (as boiling point values are not given in ref. 127) were measured in these mixtures for various concentrations from 0 to 1 mole fraction of donor and at various temperatures, with an aim to study the following :

(*a*) (*i*) The nature of velocity versus concentration curve and the influence of number of chlorine atoms present in chloroparaffin over the nature of the curve.

(*ii*) The nature of compressibility versus concentration curve and to see what light it throws over the predominence of occurrence of H-bonds over dipole association and vice-versa.

(*iii*) the behaviour of free lengh.

(*b*) Calculation of Rao's constant and Van der waal's constant (b) for the C-H......O bond.

Observations

The ultrasonic velocity, density and boiling point measured for all the mixtures at various concentrations are tabulated in tables 1 to 10. The density values are at 30° and also at 40°C (for certain concentrations) and velocity values are at various tempratures below the boiling points of components. The details about the apparatus, the method of taking observations etc. have already been mentioned in Chapter II.

ANALYSIS OF THE DATA : RESULTS AND DISCUSSION

SECTION A

Velocity in Mixtures at 30°C

Curves for the velocity versus temp. for each mixtures at each concentration were drawn. One of these, for methylene chloride-propionaldehyde, mixture, appears in fig. 3.

Velocity (Mt/Sec) and density (CM/CC) values. At various temperatures (°C) and for various concentration of Donor (mole fractions)

TABLE 1

Mixture : Methylene Chloride—Propionaldehyde

Concentration = 0.0 Mole fraction
Boiling point = 40°C
Density at 30°C = 1.280
40°C = 1.266
50°C = 1.252

Temperature	15.6	17.4	21.0	23.7	27.7	30.1	32.7	33.8
Velocity	1108.8	1103.2	1087.4	1076.0	1059.6	1049.8	1040.0	1035.0

Concentration = 0.26 Mole fractions
Boiling point = 44.8°C
Density at 30°C = 1.164

Tempreature	17	21	24.8	28	32.1	34
Velocity	1108.8	1092.0	1076.0	1063.4	1047.6	1035.6

Concentration = 0.49 Mole fraction
Boiling point = 37.2°C
Density at 30°C = 1.057
40°C = 1.046
50°C = 1.035

Temperature	19	23.8	28.8	30.7	36	37.5	38.7
Velocity	1128.6	1099.0	1080.4	1072.2	1050.4	1045	1040

Concentration = 68 Mole fraction
Boiling point = 48°C
Density at 30°C = 0.9651

Temperature	23	27	29.6	32	36.9	39.1	41
Velocity	1125	1109.6	1098.7	1089.2	1069.8	1060.8	1053.3

Concentration = 85 Mole fraction
Boiling point = 49.1°C
Density at 30°C = 0.8820

Temperature	29	32.9	36.7	38.6	40.2	43
Velocity	1127.6	1112.3	1096.4	1087.9	1081.8	1071.0

Concentration = 100 Mole fractions
Boiling Point = 48.7°C
Density at 30°C = 0.8136
40°C = 0.8046
50°C = 0.7956

Temperature	25	28	31.8	35	38.9	43.1
Velocity	1174	1161.4	1146	1130.8	1114.8	1097.7

TABLE 2

Mixture : **Chloroform—Acetones**

Concentration = 0.0
Boiling point = 61°C
Density at 30°C = 1.453
40°C = 1.434
50°C = 1.414

Temperature	26	31.6	34.8	36.8	38.8	41.4	43.8	47.9
Velocity	977.2	957.6	947.6	941.6	974.2	926.0	917.6	903.6

Concentration = 0.34 Mole fractions
Boiling point = 63.9°C
Density at 30°C = 1.227

Temperature	27.4	31.4	34.9	38.6	41.1	44.25	26	49.8
Velocity	995.2	982.3	971.4	958.4	950.4	940.4	934.6	922.0

Concentration = 0.58 Mole fractions
Boiling point = 62.3°C
Density at 30°C = 1.061
40°C = 1.049
50°C = 1.037

Temperature	26	29	32.8	36	38.8	41.6	43.5	45.4
Velocity	1032	1021.2	1008.6	997.0	987.8	977.0	971.0	965.0

Concentration = 0.75 Mole fractions
Boiling point = 60.3°C
Density at 30°C = 0.9419

Temperature	27.6	32	34.2	38.5	41.6	44.2	46.4
Velocity	1076	1059.6	1041.1	1024.8	1012.4	1002	965.5

Concentration = 0.89 Mole fractions
Boiling point = 58.5°C
Density at 30°C = 0.8399

Temperature	27	31.5	37	41.2	42.8	47.2
Velocity	1106.4	1088.0	1066.3	1049.4	1042.6	1026.0

Concentration = 1.0 Mole fractions
Boiling point = 56°C
Density at 30°C = 0.7590
40°C = 0.7541
50°C = 0.7492

Temperature	27.4	30.9	33.6	36.9	42.1	46
Velocity	1153.2	1137.6	1126	1109.6	1088.0	1070.6

TABLE 3

Mixture : **Chloroform Propionaldehyde**

Concentration = 0.2 Mole fractions
Boiling point = 61.6°C
Density at 30°C = 1.338

Temperature	26.4	31.6	34.4	38.9	42	43.7	45.5
Velocity	989.6	972.4	963.2	948.5	938.4	932.4	926.8

Concentration = 0.34 Mole fractions
Boiling point = 60.8°C

Temperature	27.8	30.7	33.2	37.1	39.2	40.8	45.6	49
Velocity	1002.0	991.8	983.8	970.2	963.1	957.8	941.8	930.4

Concentration = 0.58 Mole fractions
Boiling point = 57.8°C
Density at 30°C = 1.107
40°C = 1.096
50°C = 1.085

Temperature	26.2	32.6	37.5	38.6	42.1	42.7	45.6
Velocity	1042.0	1120.8	1002.8	1098.8	987.8	985.8	976.4

Concentration = 0.75 Mole fractions
Boiling point = 54.6°C
Density at 30°C = 0.9936

Temperature	26.9	31.7	37.2	40.4	47.4
Velocity	1080.8	1062.9	1043.5	1031.4	1006.0

Concentration = 0.89 Mole fractions
Boiling point = 52°C
Density at 30°C = 0.8948

Temperature	27	34	38.2	42.2	46.4
Velocity	1124.8	1097.8	1081.7	1065.8	1050.8

TABLE 4

Mixture : **Chloroform—Ethly Methyl Ketone**

Concentration = 0.29 Mole fractions
Boiling point = 74°C
Density at 30°C = 1.238

Temperature	26.6	31.2	32.0	35.3	36.7	38.5	41.6	45.4	46.9
Velocity	1112.4	996.7	993.4	982.8	976.8	971.6	961.2	949.9	944

Concentration = 0.52 Mole fractions
Boiling point = 77.3°C
Density at 30°C = 1.078
40°C = 1.067
50°C = 1.056

Temperature	26.6	31.5	35.8	39.3	42.3	43.8	48.7
Velocity	1052.9	1035.8	1020.8	1008.2	997.4	991.2	975.2

Concentration = 0.71 Mole fractions
Boiling point = 78.4°C
Density at 30°C = 0.9588

Temperature	25.2	31.6	33.9	37.6	41.7	43.9
Velocity	1101.6	1078.0	1069.0	1055.0	1039.4	1031.6

Concentration = 0.86 Mole fractions
Boiling point = 78.5°C
Density at 30°C = 0.8694

Temperature	27	30.7	34	37.3	42.6	45.5	47.4
Velocity	1138	1122.6	1109.2	1096.6	1075.0	1063.6	1056.2

Concentration = 0.93 Mole fractions
Boiling point = 78.5°C
Density At 30°C = 0.8287

Temperature	27	31.8	37.6	43.1	45	48.1
Velocity	1162	1141.4	1118	1095.6	1088.2	1075.4

Concentration = 1.0 Mole fractions
Boiling point = 78.5°C
Density at 30°C = 0.7811
40°C = 0.7725
50°C = 0.7639

Temperature	32.9	35.9	37.8	40.6	44.6	46.4	50
Velocity	1156.0	1143.4	1135.8	1124.4	1107.2	1100.0	1084.7

TABLE 5
Mixture : Dichloroethane—Ethyl Methyl Ketone

Concentration = 0.0
Boiling point = 82.3°C
Density at 30°C = 1.222
40°C = 1.209
50°C = 1.183

Temperature	26	32.1	34.8	36.3	38.1	41.3	44.2	49.4
Velocity	1188.4	1162.4	1152.8	1146.6	1139.4	1126.4	1115.4	1094.8

Concentration = 0.14
Density at 30°C = 1.149

Temperature	25.8	31.2	33.8	39.2	42.8	46.5
Velocity	1179.6	1158.2	1148.0	1126.6	1112.8	1098.8

Concentration = 0.35
Boiling point = 82.7°C
Density at 30°C = 1.050

Temperature	32	36	39.4	42.6	45.4	47.3
Velocity	1146.8	1131.6	1118.4	1105.6	1095.2	1088.4

Concentration = 0.25
Boiling point = 83°C
Density at 30°C = 1.114

Temperature	26	31.4	34.0	36.9	41.6	44.4	49.0
Velocity	1172	1151.4	1141.5	1130.6	1112.4	1101.6	1084.0

Concentration = 0.48
Boiling point = 82.2°C
Density at 30°C = 0.9954

Temperature	25.6	30.8	34.4	38.1	40.0	43.0	45.8	48.8
Velocity	1171.6	1151.2	1137.2	1122.4	1114.6	1102.8	1093.2	1081.6

Concentration = 0.57
Boiling point = 81.3°C
Density at 30°C = 0.9145
40°C = 0.9101
50°C = 0.9057

Temperature	25.8	31.1	34.6	38.1	40	42.5	47.2
Velocity	1172	1152	1137.6	1123.8	1116.7	1108	1088.4

Concentration = 0.84
Boiling point = 80°C
Density at 30°C = 0.8450

Temperature	26.6	32.6	35.2	41.5	44.5	49
Velocity	1174	1150.4	1140	1114.8	1102.4	1084.4

TABLE 6

Mixture : **Trichloro Ethane Dioxane**

Concentration = 0.0
Boiling polnt = 112.2°C
Density at 30°C = 1.410
40°C = 1.395
50°C = 1.380

Temperature	24.6	29.8	32.6	33.6	39.2	43.2	47.6
Velocity	1146	1128	1115	1115	1096	1082	1068

Concentration = 0.14
Boiling point = 113°C
Density at 30°C = 1.356

Temperature	30.1	32.8	32.2	41.8	45.5	47.7
Velocity	1147.3	1138.3	1118.1	1106.0	1092.9	1084.8

Concentration = 0.27
Boiling point = 113.4°C
Density at 30°C = 1.311

Temperature	26	30.7	34.5	37.6	43.9	46.4
Velocity	1183.2	1165.4	1151.3	1140.1	1117.8	1108.3

Concentration = 0.39
Density at 30°C = 1.266

Temperature	26.8	31.2	34.8	38.1	40.6	44	46.8
Velocity	1199.2	1183.2	1169.7	1157	1147.9	1136.1	1125.9

Concentration = 0.50
Boiling point = 1112°C
Density at 30°C = 1.223
40°C = 1.212
50°C = 1.201

Temperature	26	31	35.5	37.7	40.7	43.6	47.6
Velocity	1221.6	1202.4	1185	1177.2	1165.7	1154.8	1140.0

Concentration = 0.69
Boiling point = 108°C
Density at 30°C = 1.147

Temperature	31.5	35.1	38.1	42.1	45.8	49.8
Velocity	1238.6	1224.5	1213.1	1196.6	1181.7	1164.8

Concentration = 1.0
Boiling point = 101°C
Density At 30°C = 1.016
40°C = 1.007
50°C = 0.998

Temperature	28	31.4	36	39.3	43.8	46.1
Velocity	1329.8	1314.8	1295.2	1280	1268	1251.2

TABLE 7

Mixture : **Trichloroéthane—Ethyl Butyrate**

Concentration = 0.22
Boiling point = 115°C
Density at 30°C = 1.248

Temperature	25.6	31.6	34.8	35.9	37.1	39.6	40.1	40.6	44.4
Velocity	1126	1105.2	1100.7	1096.7	1092.7	1083.5	1081.5	1065.5	1080

Concentration = 0.44
Boiling point = 116.5°C
Density at 30°C = 1.116
40°C = 1.104
50°C = 1.092

Temperature	26.4	30.3	32.2	34.1	35.6	39.6	46.0
Velocity	1134	1119.5	1112.2	1105.2	1099.2	1084.5	1067.8

Concentration = 0.63
Boiling point = 118°C
Density at 30°C = 1.012

Temperature	28.6	32.6	35.5	37.5	39	39.9	43	48
Velocity	1135.8	1119.6	1108.3	1100.8	1095.0	1001.8	1078.8	1068

Concentration = 0.73
Density at 30°C = 0.9661

Temperature	26.6	29.6	33	34.8	39.5	41.4	45.4	47.8
Velocity	1148.0	1135.7	1121.5	1114.7	1096.0	1089.2	1081	1071

Concentration = 0.81
Boiling point = 119.9°C
Density at 30°C = 0.9285

Temperature	30.2	32.3	34.8	37.3	41.8	45.1	48
Velocity	1140.2	1130.4	1121	1111.2	1093.8	1081.6	1070

Concentration = 0.91
Boiling point = 119.9°C
Density at 30°C = 0.8851

Temperature	25.6	29.9	32.6	35.2	37.5	41	46
Velocity	1162.4	1145.2	1134.2	1123.6	1114.4	1080.8	1080.4

Concentration = 1.0
Boiling point = 119°C
Density at 30°C = 0.8502
40°C = 0.8382
50°C = 0.8262

Temperature	25.8	28.4	33.5	35.3	39.2	44.6
Velocity	1170	1159.5	1138.0	1130.2	1144	1092

TABLE 8

Mixture : **Tetracholoroethane-Cyclohexanone**

Concentration = 0.0
Boiling point = 146°C
Density at 30°C = 1.554
40°C = 1.503
50°C = 1.452

Temperature	27	33	36.7	39.1	40.6	43.6	47.1
Velocity	1139	1119.8	1108	1100.2	1095.9	1086.2	1075.5

Concentration = 0.42
Boiling point = 156°C
Density at 30°C = 1.305
40°C = 1.293
50°C = 1.281

Temperature	26.6	30.4	34.9	38.9	42.2	44.3	45.6
Velocity	1225.6	1213	1197.8	1184.8	1175.0	1167.2	1163.2

Concentration = 0.53
Boiling point = 158.8°C
Density at 30°C = 1.235

Temperature	27.6	30.6	33.3	40.3	43.2	48.9
Velocity	1247.6	1237.8	1228.7	1205.4	1196.6	1177.5

Concentration = 0.63
Boiling point = 158.9°C
Density at 30°C = 1.178

Temperature	25.8	29.8	35.5	38.4	40.9	43.5	48.2
Velocity	1280.0	1266.4	1246.8	1237.4	1228.8	1220.0	1204.4

Concentration = 0.68
Boiling point = 159°C
Density at 30°C = 1.137

Temperature	26.8	30.5	35	37	39.8	43.1	44.4	48.6
Velocity	1289.2	1276	1261.6	1255.2	1245.2	1234	1229.2	1214.6

Concentration = 0.72
Boiling point = 158.9°C
Density at 30°C = 1.111

Temperature	25.8	30.8	34	36.4	39.4	43.5	45.6	47.6
Velocity	1304.8	1287.4	1278	1267.3	1259.4	7244.4	1237.8	1230.2

Concentration = 0.80
Boiling point = 158°C
Density at 30°C = 1.061

Temperature	28.8	31.7	34.5	37.4	42.3	46.8	48.4
Velocity	1318.4	1307.8	1298.0	1287.7	1270.0	1254.2	1248.8

Concentration = 1.0
Boiling point = 156°C
Density at 30°C = 0.9266
40°C = 0.9170
50°C = 0.9154

Temperature	32.6	35.2	38.4	40.2	42.5	43.4	46.8
Velocity	1378	1367.2	1353.5	1345.6	1336.7	1331.2	1318

TABLE 9

Mixture : **Tetrachloroethane-Mesityl Oxide**

Concentration = 0.083
Boiling point = 147°C
Density at 30°C = 1.496

Temperature	26.8	31.6	34.2	36.9	41.2	43	44.8	48.5
Velocity	1141.2	1126.4	1118.2	1109	1096.3	1090.6	1085	1073.6

Concentration = 0.3
Boiling point = 146°C
Density at 30°C = 1.328

Temperature	23.8	29	34	41.3	45
Velocity	1163.6	1147	1131	1108.2	1096.4

Concentration = 0.53
Boiling point = 142°C
Density at 30°C = 1.166
40°C = 1.155
50°C = 1.144

Temperature	30.6	35	38	40.1	43.8	48.8
Velocity	1171.0	1156.8	1147.4	1140.6	1128	1112.8

Concentration = 0.72
Boiling point = 138°C
Density at 30°C = 1.029

Temperature	26	31.2	34.8	41.1	43.4	48.2
Velocity	1213	1196	1182.8	1161.2	1153.2	1137.0

Concentration = 0.87
Boiling point = 134.2°C
Density at 30°C = 0.9218

Temperature	29.2	31.4	37.3	41.6	45.2	49.7
Velocity	1233.8	1225.6	1204.6	1188.7	1174.7	1159.2

Concentration = 1.0
Boiling point = 130°C
Density at 30°C = 8314
40°C = 8239
50°C = 8164

Temprature	28	31.1	34.2	37.3	44.3	48.8
Velocity	1274.8	1261.4	1248.4	1235.2	1206.2	1188.0

TABLE 10

Mixture : **Pentachloroethane-Cyclohexanone**

Concentration = 0.0
Boiling point = 162°C
Density at 30° = 1.645
40° = 1.628
50° = 1.611

Temperature	31.3	33.8	37	40.1	45.3	47.3	49.3
Velocity	1062.6	1055.3	1045.5	1035.6	1020.8	1015.5	1010.5

Concentration = 0.2
Density at 30°C = 1.522

Temperatute	31.6	33.2	37.6	43	46.8	51.6
Velocity	1102.5	1097.5	1084.0	1067.6	1056.5	1042.4

Concentration = 0.34
Boiling point = 165.2°C
Density at 30°C = 1.426

Temperature	25.8	30.2	34.2	38.5	46.2	49.7
Velocity	1157.6	1143.6	1131.5	1118	1094.8	1083.2

Concentration = 0.52
Density at 30°C = 1.310
40°C = 1.303
50°C = 1.296
Concentration = 0.58
Boiling point = 164.4°C
Density at 30°C = 1.255

Temperature	27.6	31.7	39.9	43.9	46.8	48.3
Velocity	1219.0	1206.1	1179.5	1166.4	1156.9	1152.8

Concentration = 0.75
Boiling point = 162°C
Density at 30°C = 1.131

Temperature	27.6	31.5	34.6	36.6	42.6	48.6
Velocity	1283.6	1269.6	1258.0	1250.9	1228.3	1207.5

Concentration = 0.82
Boiling point = 160.7°C
Density at 30°C = 1.069

Temperature	27.2	31	34.1	37.1	41.4	44.8	49.3
Velocity	1322.8	1307.1	1295.6	1283.6	1267.9	1255.5	1238.1

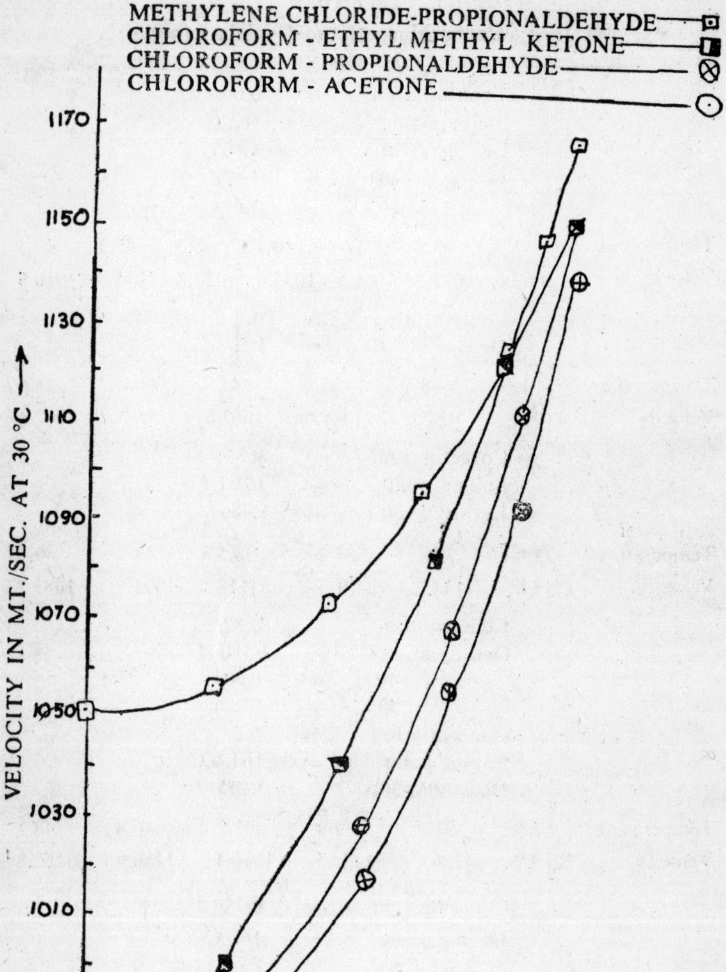
METHYLENE CHLORIDE-PROPIONALDEHYDE
CHLOROFORM - ETHYL METHYL KETONE
CHLOROFORM - PROPIONALDEHYDE
CHLOROFORM - ACETONE
1170
1150
1130
1110
1090
1070
1050
1030
1010
990
970
VELOCITY IN MT./SEC. AT 30 °C
0
0·2
0·4
0·6
0·8
1.0
MOLE FRACTION OF ELECTRON DONOR

Fig, 4

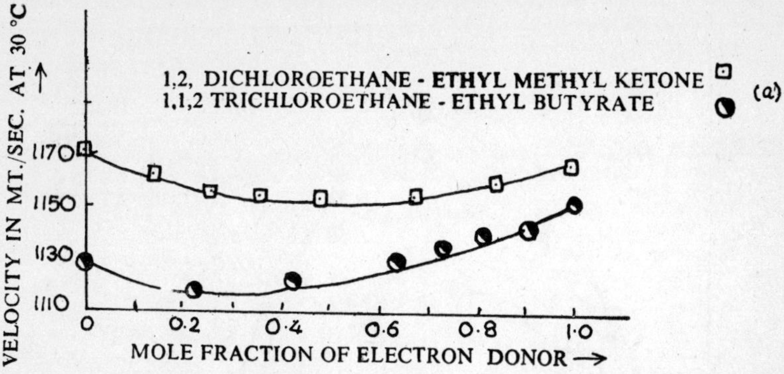

VELOCITY IN MT./SEC. AT 30 °C →
1,2, DICHLOROETHANE - ETHYL METHYL KETONE
1,1,2 TRICHLOROETHANE - ETHYL BUTYRATE
(a)
1170
1150
1130
1110
0
0.2
0.4
0.6
0.8
1.0
MOLE FRACTION OF ELECTRON DONOR →

Fig. 5 a

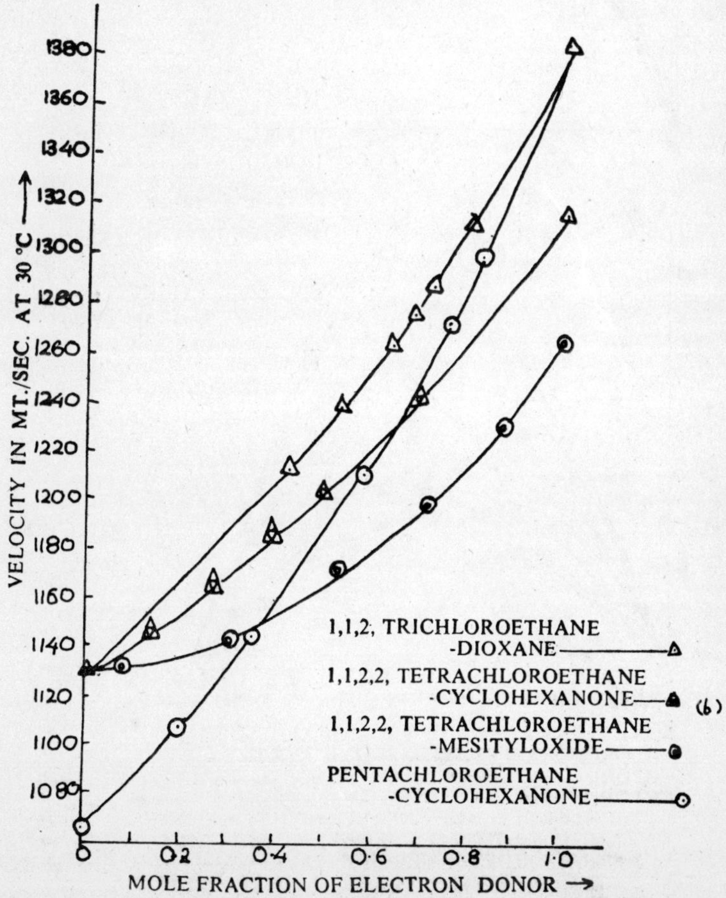

VELOCITY IN MT./SEC. AT 30 °C →
1380
1360
1340
1320
1300
1280
1260
1240
1220
1200
1180
1160
1140
1120
1100
1080
1,1,2, TRICHLOROETHANE -DIOXANE
1,1,2,2, TETRACHLOROETHANE -CYCLOHEXANONE
(b)
1,1,2,2, TETRACHLOROETHANE -MESITYLOXIDE
PENTACHLOROETHANE -CYCLOHEXANONE
0
0.2
0.4
0.6
0.8
1.0
MOLE FRACTION OF ELECTRON DONOR →

Fig. 5 b

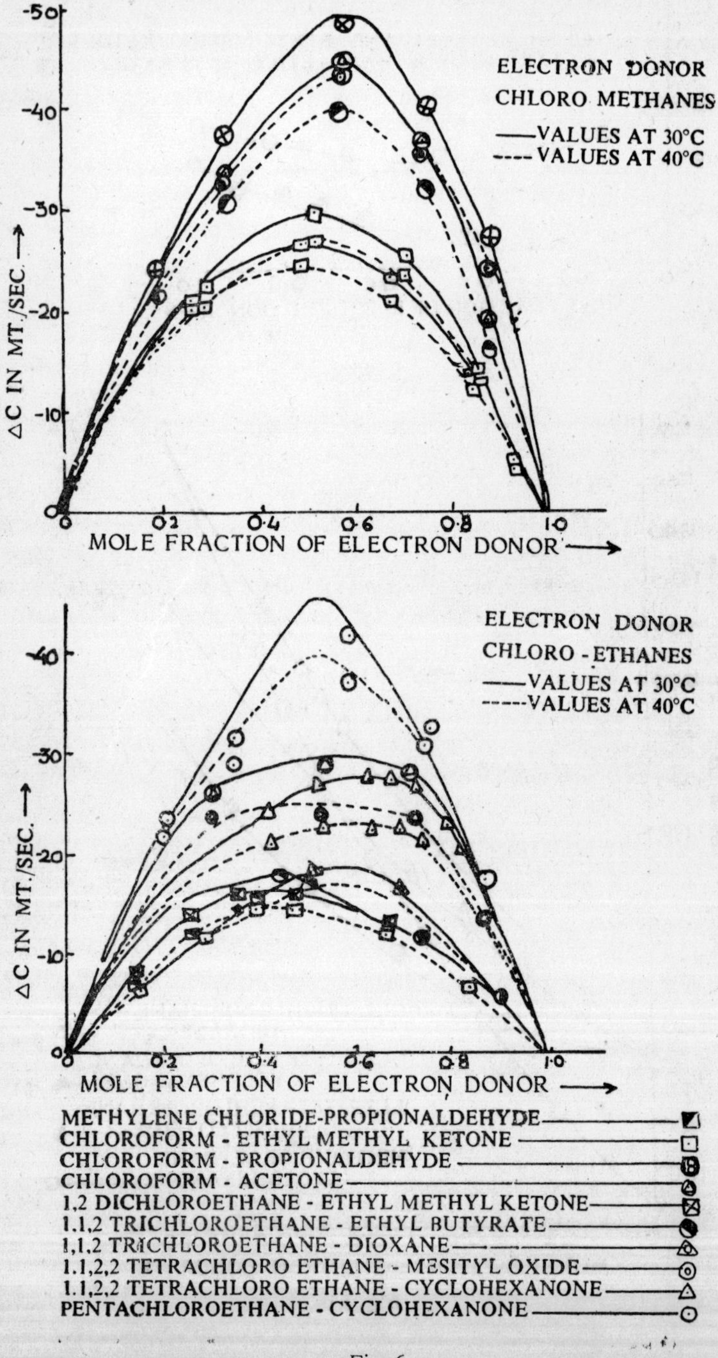
-50
-40
-30
-20
-10
0
ΔC IN MT./SEC. →
0.2
0.4
0.6
0.8
1.0
MOLE FRACTION OF ELECTRON DONOR →
ELECTRON DONOR
CHLORO METHANES
——VALUES AT 30°C
----VALUES AT 40°C
-40
-30
-20
-10
0
ΔC IN MT./SEC. →
0.2
0.4
0.6
0.8
1.0
MOLE FRACTION OF ELECTRON DONOR →
ELECTRON DONOR
CHLORO -ETHANES
——VALUES AT 30°C
-----VALUES AT 40°C
METHYLENE CHLORIDE-PROPIONALDEHYDE
CHLOROFORM - ETHYL METHYL KETONE
CHLOROFORM - PROPIONALDEHYDE
CHLOROFORM - ACETONE
1,2 DICHLOROETHANE - ETHYL METHYL KETONE
1.1,2 TRICHLOROETHANE - ETHYL BUTYRATE
1,1,2 TRICHLOROETHANE - DIOXANE
1.1,2,2 TETRACHLORO ETHANE - MESITYL OXIDE
1.1,2,2 TETRACHLORO ETHANE - CYCLOHEXANONE
PENTACHLOROETHANE - CYCLOHEXANONE

Fig. 6

Plots of velocity at 30°C versus mole fraction of donor for ten mixtures are shown in figs. 4 & 5.

From these plots, it is clear that all the mixtures undergo a negative deviation from linearity. This is similar to the observation made by Ram Prashad for mixtures chloroform-acetone and chloroform-ether. In dichloroethane-ethyl Methyl Ketone and trichloroethane-ethyl-butyrate mixtures, where the difference in velocity in the two components is small (3mt/sec. and 25 mt/sec. respectively) the velocity versus concentration curve even goes through a minima.

These velocity deviations from linear value (as read from figs. 4 & 5) for the ten mixtures are shown in fig. 6 and Table 12 and the maximum are tabulated in Table 13. A study of the plots (fig 6) show that the maximum velocity deviation depends upon the number of chlorine atoms present in the electron acceptor. Considering e.g. the chloromethanes, Table 13 shows that the maximum deviation is less in mixture with methylene chloride (which contains only two chlorine atoms) as acceptor, where as it is more in the mixture which contains chloroform (contaning chlorine three atoms) an acceptor. Similarly in chloroethanes as the number of chlorine atoms increases, the maximum deviation in velocity is also found to increase. Such increase of maximum deviation with increase in number of chlorine atoms in a paraffin is not very surprising.

TABLE 11

Velocity in Mt/Sec in mixtures at 30°C for various concentrations (In Mole fraction) of electron Donor

a. Methylenechloride—Propionaldehyde

Concentration	0	.26	.49	.68	.85	1.00
Velocity	1050.8	1056.2	1074.7	1097.1	1128.5	1153.1

b. Chloroform—Acetone

Concentration	0	.34	.58	.75	.89	1.00
Velocity	963.7	987	1018.0	1056.3	1094.4	1141.3

c. Chloroform—Propionaldehyde

Concentration	0	.2	.34	.58	.75	.89	1.00
Velocity	963.7	977.8	994.6	1028.8	1069.0	113	1153.1

d. Chloroform—Ethylmethyl Ketone

Concentration	0	.29	.52	.71	.86	.93	1.00
Velocity	963.7	1000.6	1040.8	1083.6	1125.7	1149.2	1168.6

e. 1,2 Dichloro Ethane—Ethyl Methyl Ketone

Concentration	0	.14	.25	.35	.48	.67	.84	1.00
Velocity	1171.6	1162.8	1157	1154.5	1154.2	1156	1160.8	1168.6

f. 1,1,2 Trichloroethane—Dioxane

Concentration	0	.14	.27	.39	.5	.69	1.00
Velocity	1127.2	1147.6	1168.0	1187.6	1206.0	1244.4	1321

g. 1,1,2 Trichloro Ethane—Ethyl Butyrate

Concentration	0	.22	.44	.63	.73	.81	.91	1.00
Velocity	1127.2	1117.6	1120.4	1130	1134	1140.4	1144.4	1152.4

h. 1,1,2,2 Tetrachloro Ethane—Cyclohexanone

Concentration	0	.22	.53	.63	.68	.72	.80	1.00
Velocity	1129.2	1214.2	1240.0	1265.8	1278.4	1290.4	1314.0	1389.2

i. 1,1,2,2, tetrachloro Ethane—Mesityl oxide

Concentration	0	.083	.3	.53	.72	.87	1.00
Velocity	1129.2	1131.2	1144	1173	1199.5	1230.8	1266

j. Pentachloro Ethane—Cyclohexanone

Concentration	0	.2	.34	.58	.75	0.82	1.00
Velocity	1066.4	1107.3	144.5	1211.5	1274.7	1310.8	1389.2

TABLE 12

Difference △C of velocity (In Mt/Sec) between the experimental and linear value in mixtures at various concentration (In Mole fraction of Donor at 30°C)

Methylene Chloride—Propionaldehyde						
Concentration	.26	.49	.68	.85		
	21.1	26.2	23.2	14.2		
Chloroform—Acetone						
Concentration	.34	.58	.75	.89		
	37.1	48.7	40.1	27.3		
Chloroform—Propionaldehyde						
Concentration	.2	.34	.58	.75	.89	
	23.8	33.5	44.7	36.7	19.2	
Chloroform—Ethyl Methyl Ketone						
Concentration	.29	.52	.71	.86	.93	
	22.5	29.4	25.6	14.2	5.0	
Dichloro Ethane—Ethyl Methyl Ketone						
Concentration	.14	.25	.35	.48	.67	.84
	8.4	13.8	16	15.9	13.6	8.3
1,1,2 Trichloro Ethane—Dioxane						
Concentration	.14	1.5	5.2	8.1	6.5	
	6.7	11.8	14.4	16.8	15.5	
1,1,2 trichloro Ethane—Ethyl Butyrate						
Concentration	.22	.44	.63	.73	.81	.91
	15.1	17.9	13.1	11.6	7.2	5.7
1,1,2,2 tetrachloro Ethane—Cyclohexanone						
Concentration	.42	.52	.63	.68	.72	.80
	24.2	27	27.8	27.6	27.0	23.2
1,1,2,2 tetrachloro Ethane—Mesityloxide						
Concentration	0.83	.3	.53	.72	.87	
	9.3	26.2	28.7	28.2	17.4	
Pentachloro Ethane—Cyclo hexanone						
Concentration	.2	.34	.58	.75	.62	
	23.6	31.6	42.1	33.8	20.3	

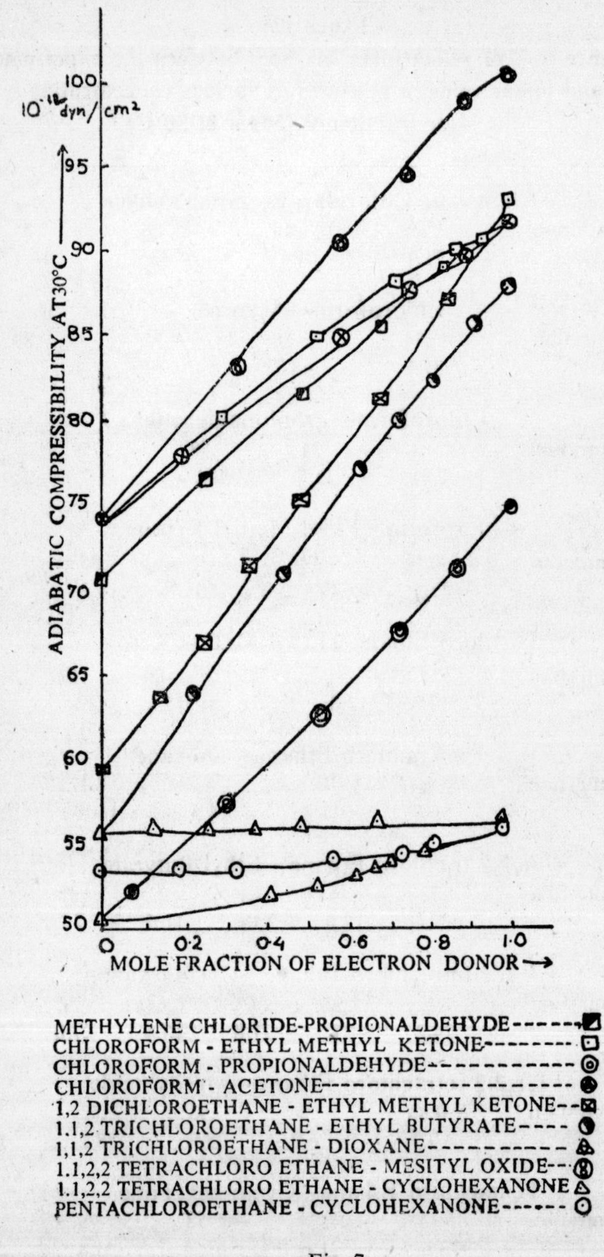
100
95
90
85
80
75
70
65
60
55
50
10^{-12} dyn/cm²
ADIABATIC COMPRESSIBILITY AT 30°C →
0
0·2
0·4
0·6
0·8
1·0
MOLE FRACTION OF ELECTRON DONOR →
METHYLENE CHLORIDE-PROPIONALDEHYDE
CHLOROFORM - ETHYL METHYL KETONE
CHLOROFORM - PROPIONALDEHYDE
CHLOROFORM - ACETONE
1,2 DICHLOROETHANE - ETHYL METHYL KETONE
1.1,2 TRICHLOROETHANE - ETHYL BUTYRATE
1,1,2 TRICHLOROETHANE - DIOXANE
1.1,2,2 TETRACHLORO ETHANE - MESITYL OXIDE
1.1,2,2 TETRACHLORO ETHANE - CYCLOHEXANONE
PENTACHLOROETHANE - CYCLOHEXANONE

Fig, 7

TABLE 13

Maximum Deviation of Velocity at 30°C

Mixture	*Concentration of donor in Mole fraction*	*Maximum deviation of velocity in Mt/Sec*
1. Methylene Chloride Propionaldehyde	.5	26.7
2. Chloroform Acetone	.6	49
3. Chloroform Propionaldehyde	.4	46
4. Chloro-Ethyl Methyl Ketone	.5	30
5. Dichloro Ethanc-Ethyl Methyl Ketone	.4	16.2
6. Trichloro ethane-Dioxane	.6	18.3
7. Trichloroethane-Ethyl butyrate	.4	18
8. 1,1,2,2 tetrachloro-ethane-cyclohexanone	.63	27.8
9. 1,1,2,2 tetrachloro-ethane Mesityl oxide	.55	30.5
10. Pentachloro ethane Cyclohexanone	.5	46

More the chlorine atoms more would be the activating effect and hence more interaction would be present. Velocity deviation thus appears to be a measure of overall interaction between the two components the interaction however may be of either kind viz: Dipole association or H_2 bonding type. As suggested by Ram Prashad, compressibility studies would be helpful in knowing the predominance of one over the other.

(*ii*) Compressibility in Mixtures at 30°C

Compressibility values at 30°C from the velocity and density data using the relation $\beta = \frac{1}{PC^2}$ are presented in Table 14. Fig. 7 shows the plot of these values against mole fraction of donor.

A glance at these plots shows that these do not behave in like manner for all the mixtures. In some of them the deviation is negative and in some it is positive. The values of maximum deviations from linearity alongwith concentration of donor liquid is given in Table 15. In chloromethanes, both the members have positive maximum deviation and its numerical value increase with increase in number of chlorine atoms. In chloro-ethanes no such regularity is noticed. Ebillioscopic study reports the presence of H_2 bonds in all the mixtures (as seen by the occurrence of maxima in boiling point versus concentration curves). The compressibility study (made by author being discussed here) reveals that the hydrogen bonds are predominantly present.

TABLE 14

Compressibility of mixtures at 30°C at various concentration (In Mole fraction) of electron donor

a. Methylene chloride—Propionaldehyde

Concentration	0	.26	.49	.68	.85	1.0
Compressibility 10^{12} sq. cm/dyne	70.73	77.05	81.93	86.12	89.80	92.47

b. Chloroform—Acetone

Concentration	0	.34	.58	.75	.89	1.0
Compressibility 10^{12} sq. cm/dyne	74.11	83.66	90.97	95.06	99.38	101.2

c. Chloroform—Propionaldehyde

Concentration	0	.2	.58	.75	.89	1,0
Compressibility 10^{12} q. scm./dyne	74.11	78.14	85.37	88.10	90.22	92.47

d. Chloroform—Ethyl Methyl Ketone

Concentration	0	.29	.52	.71	.86	.93	1.00
Compressibility 10^{12} sq. cm./dyne	74.11	80.71	85.63	88.84	90.82	91.37	93.76

e. Dichloroethane—Ethyl Methyl Ketone

Concentration	0	.14	.25	.35	.48	.67	.84	1.00
Compressibility 10^{12} sq. cm./dyne	59.64	64.33	67.07	71.48	75.44	81.87	87.76	93.76

f. Trichloroethane—Dioxane

Concentration	0	.14	.27	.39	.50	.69	1.00
Compressibility 10^{12} sq. cm./dyne	55.85	55.93	55.93	56.03	56.22	56.28	56.40

g. Trichloro Ethane Ethyl Butyrate

Concentration	0	.22	.44	.63	.73	.81	.91	1.00
Compressibility 10^{12} sq. cm./dyne	55.85	64.18	71.39	77.38	80.46	82.81	86.24	88.57

h. Tetrachloro Ethane—Cyclohexanone

Concentration	0	.42	.53	.63	.68	.72	.80	1.00
Compressibility 10^{12} sq. cm/dyne	50.49	52	52.66	52.98	53.81	54.06	54.59	55.91

i. Tetrachloroethane—Mesityl oxide

Concentration	0	.083	.3	.53	.72	.87	1.00
Compressibility 10^{12} sq. cm./dyne	50.49	52.25	57.51	62.33	67.54	71.60	75.06

j. Pentachloro Ethane—Cyclohexanone

Concentration	0	.2	.34	.58	.75	.82	1.00
Compressibility 10^{12} sq. cm./dyne	53.46	53.60	53.51	54.30	54.42	54.44	55.91

TABLE 15

Maximum deviation of Compressibility at 30°C

Mixture	*Concentration of donor (in M.F.)*	*Compressibility at 30°C (Experimental) × 10^{12} sq. cm./dyne*	*Compressibility (on linear basis) at 30°C × 10^{12} sq. cm./dyne*	*Maximum deviation of compressibility at 30°C 10^{12} sq. cm./dyne*
(1)	(2)	(3)	(4)	(5)
Methylene chloride Propionaldehyde	0.49	81.93	81.38	+ 0.55
Chloroform—Acetone	0.58	90.97	89.82	+ 1.15
Chloroform—Propionaldehyde	0.58	85.37	84.75	+ 0.62
Chloroform—Ethyl Methyl Ketone	0.52	85.63	84.32	+ 1.31
1,2 Dichloroethane Ethyl Methyl Ketone	0.67	81.87	82.50	— 0.63
1,1,2 trichloroethane Ethylbutyrate	0.44	71.39	70.24	+ 1.15
1,1,2 trichloroethane Dioxane	0.5	56.22	56.12	+ 0.10
1,1,2,2 tetrachloroethane cyclohexanone	0.42	52.00	52.76	— 0.76
1,1,2,2 tetrachloroethane Mesityl oxide	0.53	62.33	63.51	— 1.18
Pentachloroethane—Cyclohexanone	0.52	53.99	54.73	— 0.74

6 mixtures namely Methylene chloride — Propionaldehyde, Chloroform-acetone, Chloroform-propionaldehyde, Chloroform-ethyl-Methyl Ketone, trichloroethane-dioxane, trichloroethane-ethyl butyrate, show a positive deviation of compressibility from linear value, which suggests the presence of very low cohesive energy among the Molecules which is possible only of two

molecules combining together to form a single unit. In other words in all these mixtures at 30°C association due to hydrogen bond formation predominates over the dipole association. In rest of the four mixtures at 30°C, a negative deviation of compressibility from linearity is observed and hence it can be concluded that in all of them the dipole association predominates.

(*iii*) **Free length**

Experimental and linear free lengths (former obtained by the relation $L_f = K\beta^{\frac{1}{2}}$, K = 631 at 30°C) at various concentrations for the mixtures at 30°C are given in Table 16. In all the mixtures with Methylenechloride, chloroform, dichloroethane and trichloroethane as acceptors the deviations in free lengths are positive at the concentrations mentioned in Table 15. This result supports the earlier conclusion drawn from the compressibility studies viz. in all these mixtures at 30°C hydrogen bonding predominates over the dipole association. The case of mixture with dichloroethane at acceptor only seems to be an exception in which case although $\triangle\beta$ is slightly negative, the value of $\triangle Lf$ is found to be slightly positive.

TABLE 16

Experimental and theoretical (Linear) free lengths (A°) in mixture at 30°C for various concentration (in mole fraction) of electron Donor

a. Methylenechloride—Propionaldehyde

Concentration	0	.26	.49	.68	.85	1.0
Free length expt.	.5307	.5539	.5710	.5855	.5978	.6067
Free length linear	.5307	.5504	.5679	.5923	.5953	.6067

b Chloroform—Acetone

Concentration	0	.34	.58	.75	.89	1.0
Free length expt.	.5431	.57711	.6017	.6152	.6290	.6348
Free length linear	.5431	.5742	.5962	.6118	.6247	.6348

c. Chloroform—Propionaldehyde

Concentration	0	.2	.58	.75	.89	1.0
Free length expt.	.5431	.5577	.5829	.5923	.5992	.6067
Free length linear	.5431	.5558	.5799	.5908	.5997	.6067

d. Chloroform—Ethyl Methyl Ketone

Concentration	0	.29	.52	.71	.86	.93	1.0
Free length expt.	.5431	.5667	.5838	.5947	.6013	.6032	.6109
Free length linear	.5431	.5627	.5783	.5912	.6014	.6061	.6109

e. 1,2 Dichloroethane ethyl Methyl Ketone

Concentration	0	.14	.35	.48	.67	.84	1.0
Free length expt.	.4872	.5059	.5334	.5480	.5710	.5910	.6109
Free length linear	.4872	.5045	.5305	.5465	.5700	.5911	.6109

f. 1,1,2 Trichloro ethane Dioxane

Concentration	0	.14	.39	.50	.69	1.0
Free length expt.	.4715	.4719	.4723	.4731	.4734	.4739
Free length linear	.4715	.4718	.4724	.4727	.4731	.4739

g. 1,1,2 Trichloroethane—Ethyl Butyrate

Concentration	0	.22	.44	.63	.73	.81	.91	1.0
Free length expt.	.4715	.5055	.5331	.5550	.5659	.5741	.5859	.5937
Free length linear	.4715	.4983	.5252	.5454	.5607	.5704	.5827	.5937

h. 1,1,2,2 tetrachloro Ethane—Cyclohexanone

Concentration	0	0.42	.053	0.63	0.68	0.72	0.80	1.00
Free length expt.	.4483	.4550	.4578	4592	4628	.4638	.4661	.4718
Free length linear	.4483	.4581	.4607	.4631	.4642	.4652	.4671	.4718

i. 1,1,2,2 Tetrachloroethane—Masityloxide

Concentration	0	0.083	0.53	0.72	0.87	1.00
Free length expt.	.4483	.4560	.4980	.5186	.5338	.5467
Free length linear	.4483	.4564	.5004	.5191	.5338	.5467

j. Pentachloroethane—cyclohexanone

Concentration	0	.2	.34	.58	.75	.82	1.00
Free length expt.	.4613	.4619	.4615	.4649	.4654	.4655	.4718
Free length linear	.4613	.4634	.4648	.4673	.4691	.4699	.4718

In other three mixtures $\triangle Lf$ is negative suggesting the predominance of dipole association over the hydrogen bond formation. This again supports the conclusions arrived at from the compressibility studies.

Conclusion

(1) Analysis of Ultrasonic velocity data alone, for mixtures at various concentration, does not show whether in a mixture dipole association exists among the molecules of components or hydrogen bonds are present, since the infiuence of both the interactions is to make the velocity versus concentration curve undergo a negative deviation.

(2) The maximum negative deviation from the linear value of velocity has been found to be a measure of the strength of the overall interaction which has been found to increase with increase in number of chlorine atoms in a chloroparaffin.

(3) Compressibility study of the ten mixtures shows that not all the acceptor molecules are hydrogen bonded and that there exists as well, the dipole association. The compressibility versus concentration curve undergoes a positive deviation when the interaction among the hydrogen bonded molecules predominates, and a negative deviation when dipole-association predominates.

(4) At 30°C, in all the mixtures except in those with acceptors, tetra and pentachloroethane, the cohesive energy is mainly due to interaction among the hydrogen bonded molecules in rest of the mixtures dipole association predominates.

(5) The experimental free length has been found to be more than the linear value for the first seven mixtures including the one with dichloroethane as acceptor. This confirms the predominance of hydrogen bonds in all these mixtures. In mixtures with acceptors tetrachloroethane and pentachloroethane, the case is opposite and hence the presence of dipole association seems to be more prominent.

SECTION B

Van der waal's constant 'b' and Rao's constant 'R'

There has been extensive work on molar refraction values of

hydrogen bonds by Batsanov[100] and many other. Recently Puranic[94], et al made a refractometic study of hydrogen bonds in mixtures of water and methyl alcohol, with acetone, dioxane and Pyridine, the molar refraction values for O—H———O bond were calculated and were found to be of the same order as those arrived at by Batsanov[95] et al. No such study exists for Van der waal's constant b and Rao's constant R for hydrogen bonds. As the constants 'b' and 'R' have significance similar to molar refraction, it should be possible to compute the values of these constants for hydrogen bonds. These were therefore, computed for the bond CH———O by the author for all those mixtures in which compressibility study showed the predominance of hydrogen bonds. An assumption was made that the deviation observed in the velocity versus concentration curve is entirely due to the presence of hydrogen bonds, i.e. the decrease in the velocity of the mixture is due to the decrease in the velocity of the electron acceptor alone, the procedure followed for these computations is as follows:

In the absence of any interaction between the components of the mixture, velocity of the mixture will be given by :

$$C = C_1X_1 + C_2X_2 = C_1\,(1-X_1) + C_2X_2$$

$$C = C_1 + (C_2-C_1)\ X_2 \qquad \ldots(41)$$

Where X_2 is mole fraction of donor and C_1, C_2 are the velocities of acceptor and donor respectively. Now, when hydrogen bonds are present, a departure of the velocity of mixture from one given by the above relation will be there. In section A of this chapter it was noted that in all the six mixtures (that is mixtures with chloromethane, and trichloro-ethane as acceptor) where compressibility deviation is positive, or negative deviation of velocity versus concentration occurs. It is also shown in section A that the extent of the departure of the experimental curve from a linear plot (of velocity versus concentration) can be considered as a measure of strength of the interaction. Thus, if the reason for decrease in velocity of the mixture be taken as due to decrease in the velocity of the electron acceptor alone (due to formation of hydrogen bonds), the difference $\triangle c$ between the observed velocity and that predicted by above relation at a given concentration X_2 would be maximum when all of these molecules are hydrogen bonded. The velocity of the mixture C at concentration X_2 would be :

$$C^* = C_1^* + (C_2 - C_1^*) X_2 \qquad ...(42)$$

Where C_1 is the velocity of the hydrogen bonded electron acceptor.

Therefore, the maximum difference, $\triangle c$ max. between the experimental and linear expectation value occurring at concentration X_2 of donor would be given by

$$\triangle c \text{ max} = C - C^*$$

$$\triangle c \text{ max} = C_1 + (C_2 - C_1)^{\times}{}_2 - [C_1^{\times} + (C_2 - C_1^{\times})X_2]$$

$$\triangle c \text{ max} = C_1 + C_2X_2 - C_1X_2 - C_1^* - C_2X_2 + C_1^*X_2$$

$$= C_1(1 - X_2) - C_1^{\times}(1 - X_2)$$

$$= C_1X_1 - C_1^{\times}X_1 = (C_1 - C_1^{\times})X_1$$

$$\frac{\triangle c \text{ max}}{X_1} = C_1 - C_1^*$$

$$C_1^{\times} = C_1 - \frac{\triangle \text{ max}}{X_1} \qquad ...(43)$$

Thus, knowing the value of velocity of hydrogen bonded acceptor C_1^* the value of Van der waal's constant b for the hydrogen bonded acceptor can be calculated from the relation

$$h^1 = \frac{M}{\rho}\left[1 - \frac{RT}{Mc_1} X_2 \left\{\left(1 + \frac{MC_1^{\times 2}}{3\,RT}\right) - 1\right\}\right] \qquad ...(44)$$

Percent increase over normal of b $= \left(\frac{h - h^1}{h}\right) 100$. Similarly

Rao's constant 'R' for the hydrogen bonded acceptor can be calculated using the relation :

$$R' = \left(\frac{M}{\rho} C_1{}^{\frac{1}{3}*}\right) \qquad ...(45)$$

Percent increase over normal of, R

$$= \left(\frac{R - R'}{R} \right) 100 \qquad ...(46)$$

An idea of strength of CH- - -O bond can be had by comparing the b and R values for CH- - -O bond with those for CH- - -O bond (Present in water). Van der waal's constant and Rao's constant for CH- - -O bonds present in water can be calculated in the following manner Had water existed completely in close packed state (i.e. no hydrogen bonds among its molecules), the velocity at ordinary temperature, can easily be computed.

Thus, knowing the velocity C in unhydrogen bonded water (from above relation) and that in hydrogen bonded water (from the standard tables), Van der waal's constant and Rao's constant for CH- - -O bond can be computed.

Discussion and Conclusion

The values of 'b' and 'R' for the CH - - -O bond in the six mixtures, thus determined for the first time, appear in Talbe 17. Percent increase over normal of b & R for CH- - -O bond are also given alongwith these values for CH- - -O bond for water calculated by the author using the data given in the work of Subrahmanyam and Raghunath[101] for velocity of hydrogen bonded and unhydrogen bonded water at 30°C.

From Table 17, one finds that $b' < b$ & $R' < R$. This means that the hydrogen bonds make a negative contribution to the Van der waals's constant b and Rao's constant R. such negative values of Rao's constant were observed by Padmini[102] in semipolar bonds also.

HCH- - -O and RCH- - -O values have been found to decrease (that is becoming more negative) with an increase in number of chlorine atoms in chloromethanes.

Percent increase over normal of b & R for CH- - -O and CH- - -O bonds shows that CH- - -O bonds are very weak in comparison to CH- - -O bonds (Present in water). This is obvious since water is the one most highly associated of all the llquids.

TABLE 17

Van der waal's constant and Rao's constant for Hydrogen Bonds

Mixture	X_1	max. at 30°C	C_1*	Vauder Waal's constant b: Normal b	hydrogen bonded b'	Diffe-rence	% bCH-O	Rao's constant: Normal R	hydrogen bonded R'	Diffe-rence	% RCH-O
Methylene chloride Propiondeldehy	.5	26.75	997.5	61.60	61.42	—0.18	—0.29	674.7	663.0	—11.7	—1.7
Chloroform-acetone	.4	49	841.7	76.65	76.05	—0.60	—0.78	811.6	775.6	—36.0	—4.4
Chloroform Propionaldehyde	.6	46	845.7	76.65	76.08	—0.57	—0.74	811.6	777.8	—33.8	—4.1
Chloroform-ethyl Methyl Ketone	.5	30	903.7	76.65	76.37	—0.28	—0.36	811.6	794.3	—17.3	—2.1
Dichloroethane-ethyl Methyl Ketone	.6	16.2	1144.6	75.91	75.84	—0.07	—0.92	853.3	846.8	— 6.5	—0.76
Trichloroethane-Dioxane	.4	18.3	1081.5	89.23	89.04	—0.19	—0.21	984.7	970.9	—13.8	—0.7
Trichloroethane-ethyl butyrate	.6	18.0	1081.5	89.23	89.10	—0.13	—0.14	984.7	975.9	— 8.8	.89

Value of b for CH — — — O in water = — 1.8%
Value of R for CH — — — O in water = —10.4%

PART II

SECTION A

EFFECT of change in temperature in mixtures of Chloroparaffins and Electron Donor liquids on ultrasonic velocity and Compressibility :—

In the previous chapter it has been observed that in all the ten mixtures at 30°C with Chloroparaffin as electron acceptor, both the interactions viz. dipole-association as well as formation of hydrogen bonds are present. Both these have a similar influence over the deviation from linearity of the velocity versus concentration curve. Further the over all interaction and hence the maximum deviation increases with an increase in number of chlorine atoms in acceptor (belonging to a particular chloro-paraffin series). Whereas the compressibility and free length versus concentration curves undergo a positive deviation from linearity for mixtures where hydrogen bonds are present in predominance (like the two chloromethanes, trichloroethane at 30°C), there is negative deviation when dipole-association is present in predominance (like rest of the chloroethanes at 30°C). These conclusions have been arrived at when the temp. of the mixture is 30°C: A change in temp. is very much likely to change the magnitude and sign of these velocity and compressibility deviations. A study of these deviations at various temperatures would be helpful in knowing how the overall interaction changes (from velocity deviation study) and also how the two interactions compete with each other (from compressibility deviation study) as the temp. is varied. Velocity was therefore measured at various temp. and for different concentrations for all the ten mixtures. The density was measured at various temp. for all the mixtures for three concentrations of donor viz. (i) Concentration-zero (ii) Concentration-one and (iii) that concentration at which compressibility deviation was found to be maximum when the temperature was 30°C.

The aim of these measurements being (i) to see the effect of temp. on deviation in case of velocity versus concentration curve and hence on the overall interaction in these systems (ii) to see the effect of temperature on the maximum deviation in case of compressibility versus concentration curve and hence on the two interactions viz (a) one due to hydrogen bond and (b) that due to dipole-association.

The velocity and density data has already been given in Tables 1 to 10 (Chapter III).

Results and Discussions

(*i*) *Influence of temperature of Velocity Deviation*

Velocity values at 40°C given in Table 18, have been obtained from the velocity (given in Tables 1 to 10) versus temp. curves. These velocities thus, obtained at 40°C were found to be less than the linear values. These negative deviations at 40°C have been plotted alongwith those at 30°C (these appearing as dashed and continuous lines respectively) in fig. 6. The velocity deviations at 30°C and 40°C appear in Table 19. A study of the curves (fig. 6) shows that as the temp, rises, the maximum value of negative deviation from linearity decreases. In other words, the influence of increase of temp. in all the ten mixtures is to reduce the overall interaction in them. Regarding the influence of number of chlorine atoms in acceptor over maximum deviation, it appears that the increase of temp. does not change the order in chloromethane, that is, at 40°C also the maximum negative deviation increases with an increase in number of chlorine atoms in chloromethane.

(*ii*) *Effect of temp. on compressibility deviation*

The compressibility was calculated at 30°C, 40°C and 50°C at three concentrations of donor viz: (i) zero (ii) one and (iii) that concentration (say X) at which compressibility deviation is maximum at 30°C.

The deviations of the compressibility values from the linear ones at the concentration 'X' of donor were determined.

TABLE 18

Velocity in mt/sec in mixtures at 40°C for various concentrations (in mole fraction) of electron Donor

a. Methylene chloride—Propionaldehyde

Concentration	0	0.26	0.49	0.68	0.85	1.00
Velocity	1010.1	1016.1	1034.8	1057.2	1082.7	1110

b. Chloroform—Acetone

Concentration	0	0.34	0.58	0.75	0.89	1.00
Velocity	930.1	954.1	983.2	1019.1	1054.2	1096.9

c. Chloroform—Propionaldehyde

Concentration	0	0.2	0.34	0.58	0.75	0.89	1.00
Velocity	930.1	944.8	960.7	994.8	1032.8	1074.8	1110.0

d. Chloroform—Ethyl Methyl Ketone

Concentration	0	0.29	0.52	0.71	0.86	0.93	1.00
Velocity	930.1	966.5	1005.4	1046.1	1085.6	1108.4	1126.7

e. 1,2 Dichloroethane—Ethyl Methyl Ketone

Concentration	0	0.14	0.25	0.35	0.48	0.67	0.84	1.00
Velocity	1131.9	1124	1118.4	1116	1115	1117.1	1120.6	1126.0

f. 1,1,2 trichloroethane—Dioxane

Concentration	0	0.14	0.27	0.39	0.5	0.69	1.00
Velocity	1093.4	1112.4	1131.2	1150.7	1168.5	1204.8	1277.7

g. 1,1,2 trichloroethane—Ethyl Butyrate

Concentration	0	0.22	0.44	0.63	0.73	0.81	0.91	1.00
Velocity	1093.4	1082.1	1082.9	1091.2	1094.8	1101.0	1104.6	1111.1

h. 1,1,2,2 tetrachloroethane—Cyclohexanone

Concentration	0	0.42	0.53	0.63	0.68	0.72	0.80	1.00
Velocity	1097.9	1181.4	1206.8	1232.0	1244.4	1256.0	1278.4	1346.6

i. 1,1,2,2 tetrachloroethane—Mesityloxide

Concentration	0	0.083	0.30	0.53	0.72	0.87	1.00
Velocity	1097.9	1099.8	1112	1141.2	1165.1	1194.2	1224.4

j. Pentachloroethane—Cyclohexanone

Concentration	0	0.20	0.34	0.58	0.75	0.82	1.00
Velocity	1036.5	1077	1113.3	1179.0	1238.4	1273.2	1346.6

TABLE 19

Difference $\triangle c$ of velocity (in mt/sec) between the experimental and linear value in mixtures at various concentrations (In Mole fraction of Donor) at 30°C & 40°C

a. Methylenechloride—Propionaldehyde

Concentration	0.26	0.49	0.68	0.85
30°C	21.1	26.2	23.2	14.2
40°C	19.1	24.1	20.8	12.3

b. Chloroform—Acetone

Concentration	0.34	0.58	0.75	0.89
30°C	37.1	48.7	40.1	27.3
40°C	32.7	43.6	36.0	24.3

c. Chloroform—Propionaldehyde

Concentration	0.20	0.34	0.58	0.75	0.89
30°C	23.8	33.5	44.7	36.7	19.2
40°C	21.4	30.7	40.0	32.6	15.9

d. Chloroform—Ethyl Methyl Ketone

Concentration	0.29	0.52	0.71	0.86	0.93
30°C	22.5	29.4	25.6	14.2	5.0
40°C	20.6	26.9	23.6	13.5	4.5

c. Dichloroethane—Ethyl Methyl Ketone

Concentration	0.14	0.25	0.35	0.48	0.67	0.84
30°C	8.4	13.8	16.0	15.9	13.6	8.3
40°C	7.2	12.2	14.1	14.4	11.8	6.9

f. 1,1.2 trichloroethane—Dioxane

Concentration	0.14	0.27	0.39	0.5	0.69
30°C	6.7	11.5	15.2	18.1	16.5
40°C	6.7	11.8	14.4	16.8	15.5

g. 1,1,2 trichloroethane—Ethyl Butyrate

Concentration	0.22	0.44	0.63	0.73	0.81	0.91
30°C	15.1	17.9	13.1	11.6	7.2	5.7
40°C	15.2	18.0	13.4	11.6	6.8	5.0

h. 1,1,2,2 tetrachloro Ethane—Cyclohexanone

Concentration	0.42	0.53	0.63	0.68	0.72	0.80
30°C	24.2	27.0	27.8	27.6	27.0	23.2
40°C	21.0	22.9	22.6	22.7	21.1	18.6

i. 1,1,2,2 tetrachloroethane—Mesityloxide

Concentration	0.083	0.3	0.53	0.72	0.87
30°C	9.3	26.2	28.7	28.2	17.4
40°C	8.4	23.7	23.9	23.6	13.5

j. Pentachloroethane Cyclohexanone

Concentration	0.2	0.34	0.58	0.75	0.82
30°C	23.6	31.6	42.1	33.8	20.3
40°C	21.8	28.9	37.6	30.9	17.8

From the table, 15/20 one observes that as the temp. rises, the value of compressibility deviation decreases e.g. mixtures with acceptors chloromethanes and trichloroethanes having a positive deviation in them at 30°C become less at 40°C and even become negative in some of them at 50°C. Where as in mixtures with acceptors Di, tetra and Pentachloro-ethane in which the deviations are negative at 30°C become more and more negative as the temp. is increased to 40°C and to 50°C. This shows that as the temp. rises, hydrogen bonds become weaker and get broken. As a result of this the dipole association gets predominant, thus reducing the compressibility deviation.

TABLE 20

Maximum Deviation of Compressibility At 30°C 40°C & 50°C

Mixture	Concentration of donor in m.f	Compressibility at 30°C Expt.	Compressibility at 30°C Theo.	at 30°C Max.	Compressibility at 40°C Expt.	Compressibility at 40°C Theo.	at 40°C Max.	Compressibility at 50°C Expt.	Compressibility at 50°C Theo.	at 50°C Max.
1. Methylene chloride Propionaldehyde	.49	81.93	81.38	+0.55	89.25	88.94	+0.31	97.61	97.39	+0.22
2. Chloro Acetone	.58	90.97	89.82	+1.15	98.58	97.78	+0.80	107.2	106.8	+0.4
3. Chloro-Propionaldehyde	.58	85.37	84.75	+0.62	92.21	92.38	—0.17	99.82	100.94	—1.12
4. Chloro-Ethyl Methyl Ketone	.52	85.63	84.32	+1.31	92.74	91.79	+0.95	100.7	100.1	+0.6
5. Dichloroethane-Ethyl Methyl Ketone	.67	81.87	82.50	—0.63	88.10	89.70	—1.60	95.04	98.03	—2.99
6. Trichloroethane-Dioxane	.50	56.22	56.12	+0.10	60.43	60.41	+0.02	65.09	65.17	—0.08
7. Trichloroethane-Ethyl butyrate	.44	71.39	70.24	+1.15	77.23	76.11	+1.12	83.79	82.72	+1.17
8. Tetrachloroethane-Cyclohexanone	.42	52.00	52.76	—0.76	55.42	57.26	—1.84	59.14	62.11	—2.97
9. Tetrachloroethane Mesityl Oxide	.53	62.33	63.51	—1.18	66.51	68.82	—2.31	71.01	74.84	—3.83
10. Pentachloroethane-cyclohexanone	.52	53.99	54.73	—0.74	57.55	58.71	—1.36	60.92	62.83	—1.91

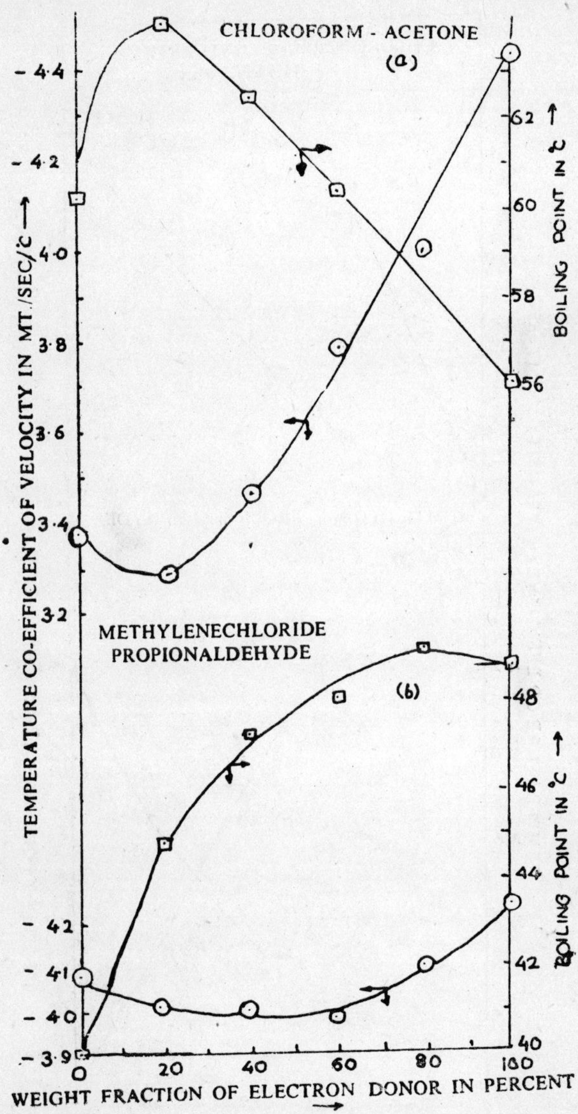
CHLOROFORM - ACETONE
(a)
METHYLENECHLORIDE
PROPIONALDEHYDE
(b)
TEMPERATURE CO-EFFICIENT OF VELOCITY IN MT./SEC/°C →
BOILING POINT IN °C ↑
BOILING POINT IN °C ↑
WEIGHT FRACTION OF ELECTRON DONOR IN PERCENT

Fig. 8

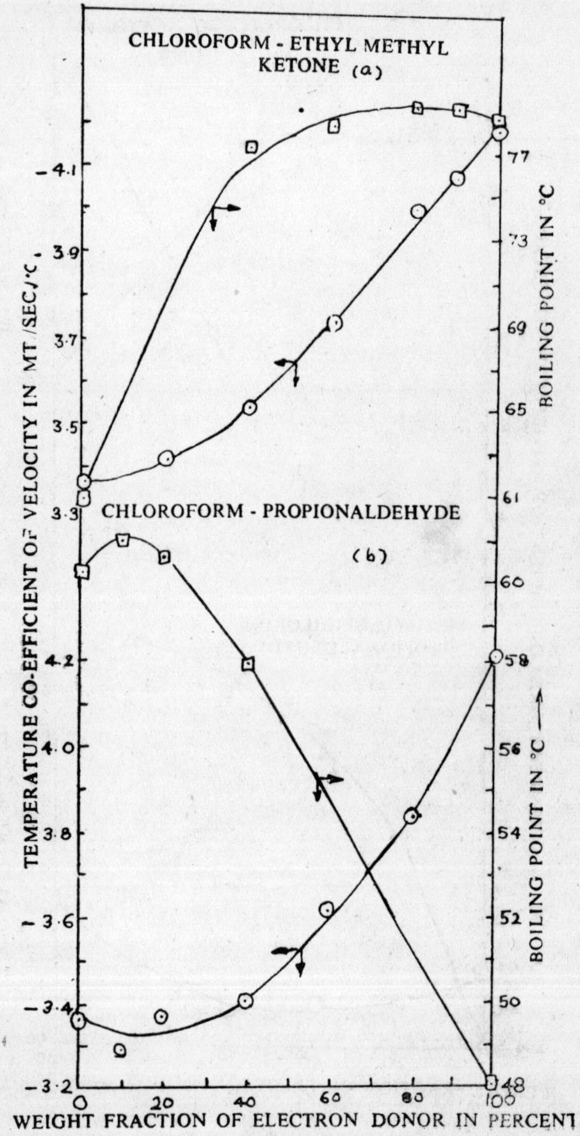
CHLOROFORM - ETHYL METHYL KETONE (a)
CHLOROFORM - PROPIONALDEHYDE
(b)
TEMPERATURE CO-EFFICIENT OF VELOCITY IN MT./SEC/°C.
BOILING POINT IN °C
WEIGHT FRACTION OF ELECTRON DONOR IN PERCENT
-4.1
3.9
3.7
3.5
3.3
4.2
4.0
3.8
-3.6
-3.4
-3.2
77
73
69
65
61
60
58
56
54
52
50
48
0
20
40
60
80
100

Fig. 9

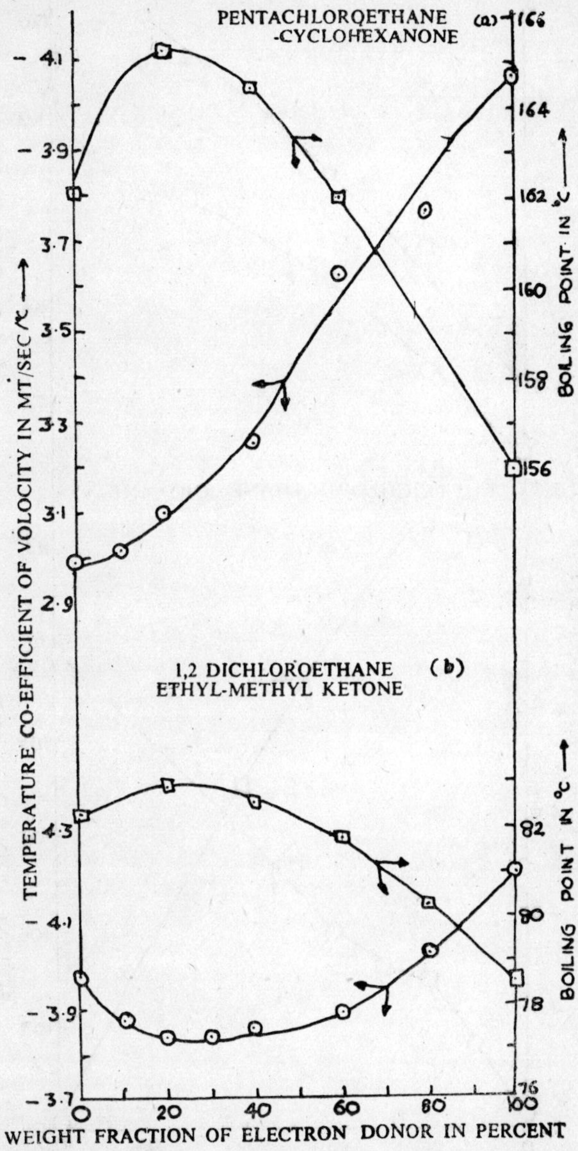
PENTACHLOROETHANE
-CYCLOHEXANONE
(a)
1,2 DICHLOROETHANE
ETHYL-METHYL KETONE
(b)
TEMPERATURE CO-EFFICIENT OF VOLOCITY IN MT/SEC/°C
BOILING POINT IN °C
WEIGHT FRACTION OF ELECTRON DONOR IN PERCENT

Fig 10

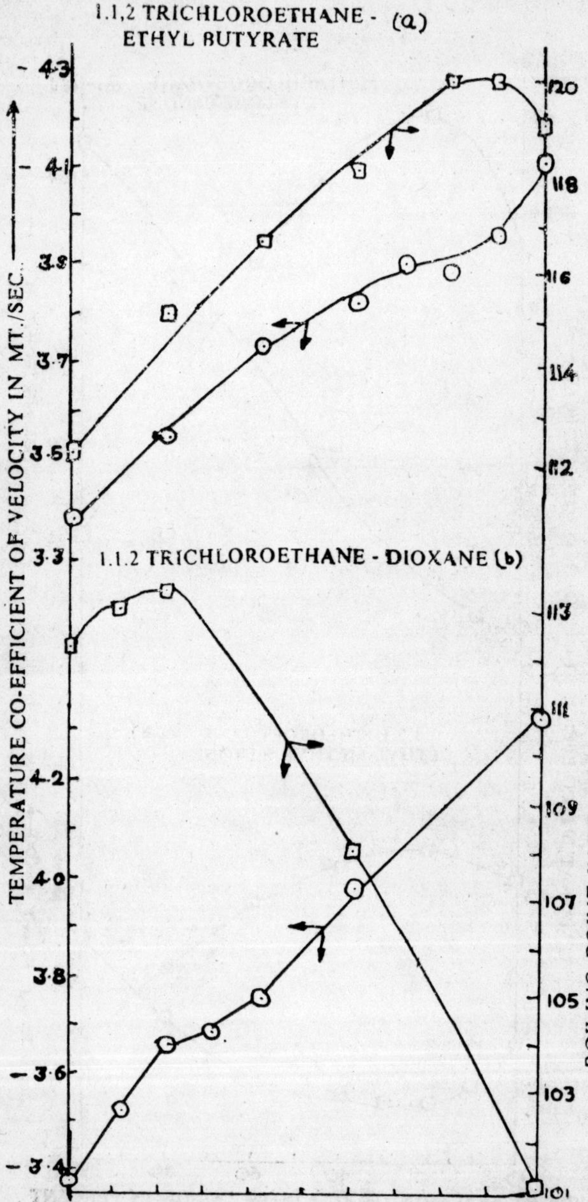

1.1,2 TRICHLOROETHANE - (a)
ETHYL BUTYRATE
1.1.2 TRICHLOROETHANE - DIOXANE (b)
TEMPERATURE CO-EFFICIENT OF VELOCITY IN MT./SEC.
BOILING POINT IN °C
WEIGHT FRACTION OF ELECTRON DONOR IN PERCENT

Fig. 11

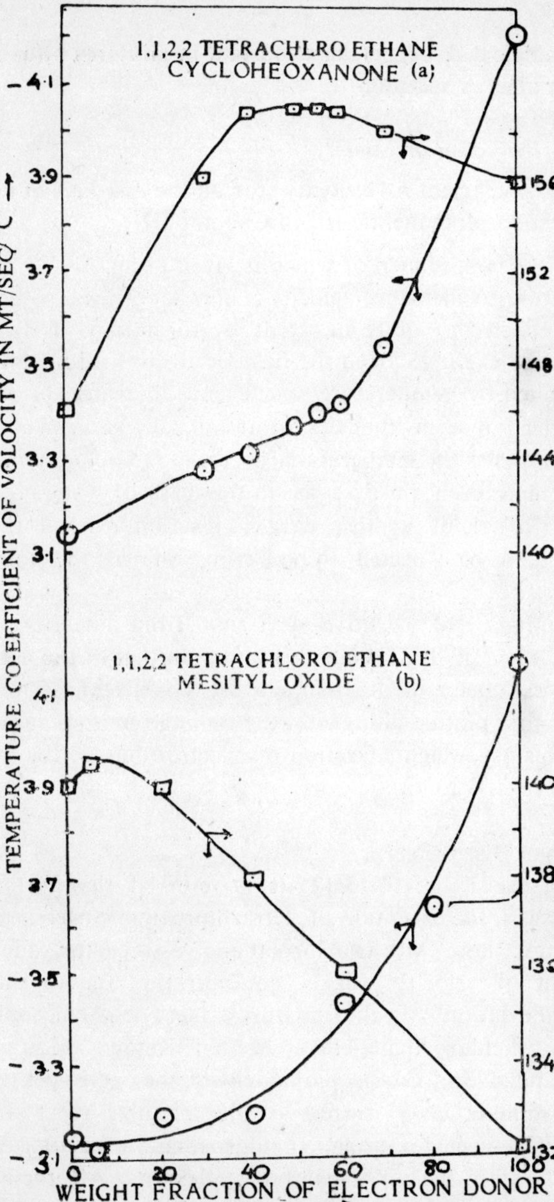
1,1,2,2 TETRACHLRO ETHANE
CYCLOHEOXANONE (a)
1,1,2,2 TETRACHLORO ETHANE
MESITYL OXIDE (b)
TEMPERATURE CO-EFFICIENT OF VOLOCITY IN MT/SEC/°C
BOILING POINT IN °C
WEIGHT FRACTION OF ELECTRON DONOR IN PERCENT

Fig. 12

SECTION B

temp. coefficient & Legemann's constant in mixtures with chloroparaffins as acceptor

(*i*) *temp. coefficient of velocity*

Temp. coefficient of velocity for all the mixtures at various concentration plots mentioned in Chapter III.

(a specimen of which is given in fig. 3)

Temperature coefficient of velocity is another of these parameters which is effected to quite an extent by formation of hydrogen bonds. For example, from the behaviour (Ref. Mc. Dade[103] et al) exhibited by temperature coefficient in water of various temperatues, it seems that the formation of hydrogen bond in a systrem malles the temperature coefficient of velocity in it less negative and even positive as in the case of water. Hence the plot of dc/dt against weight fraction for all the ten mixtures, can be expected to be having one of the following characteristics :

(a) a minima (b) negative deviation from linearity or (c) inflexion when dc/dt of donor is more than that of the acceptor.

The parameter dc/dt (temperature coefficient of velocity(was therefore plotted alongwith, the boiling point (as measured) values against weight fraction of electron donor for all the mixtures (figs. 8 to 12).

Results and Discussions

From the figures 8 to 12, it is observed that in all the mixtures with the exception of tetrachloroethane-cyclohexanone mixture and those with trichloroethane as acceptor, the temp. coefficient of velocity versus concentration curve shows a negative deviation. In the mixtures methylene-chloropropion aldehyde, dichloroethane-ethyl Methyl Ketone, chloropropin-acetone and chloroform-propionaldehyde, the curve goes through a small minima also. In rest of the mixtures (that is tetra-chloroethane-cyclohexanone, trichloroethane with dioxane and ethyl butyrate), the curve exhibits inflexion. All these three kinds (i.e. negative deviation, minima, inflexion) of the dc/dt

TABLE 21

Nature of the temperature coefficient of Velocity versus concentration and boiling point versus concentration

	Mixture	*Nature of the temperature coefficient of velocity versus concentration (in percentage weight fraction curve)*	*Concentration at which unusuality occurs*	*Concentration at which boiling point is maximum*
1.	Methylene chloride propionaldehyde	The curve goes through a minima at concentration of donor mentioned in the next column	60%	80%
2.	Chloroform Acetone	goes through a minima	20%	20%
3.	Chloroform Propionaldehyde	goes through a flat minima	10% to 20%	10%
4.	Chloroform Ethyl Methyl Ketone	Undergoes a negative deviation from linearity	—	Flat maxima between 80% & 100%
5.	1,2 Dichloro Ethane Ethyl Methyl Ketone	goes through a minima	10%	20%
6.	Trichloroethane-Dioxane	Undergoes inflexion	20% to 30%	20%
7.	Trichloroethane-Ethyl butyrate	Undergoes inflexion	70% to 90%	80% to 90%
8.	Tetrachloroethane cyclohexanone	Under goes a negative deviation from linearity. The curve abruptly changes slope at the concentration mentioned in the adjoint col. Slight inflexion at the same concentration is also present.	55%	55%
9.	Tetrachloroethane-Mesityl oxide	goes through a negative deviation and is flat for the low concentration of electron donor	0 to 10%	5%
10.	Pentachloro-ethane cyclohexanone	goes through a negative deviation	—	10%

versus concentration curve arise due to temp. coefficient of velocity of mixture becoming less negative in comparison to the linear value. Thus, the presence of hydrogen bonds in all the mixtures is indicated by temperature coefficient of velocity study also.

Comparing the temp. coefficient of velocity versus concentration curves (figs. 8 to 12 and Table 21) with the boiling point versus concentration curves (which exhibit maxima), it is found that in general the unusuality exhibited by former curves fall near about the same concentration as the one at which the boiling point is maximum.

For example :

(*a*) The Inflexion in the curve in dc/dt versus weight fraction plots (fig. 11a, 11b, 12a) for the mixtures trichloroethane with ethyl butyrate, dioxane and that of tetrachloroethane with cyclohexanone occurs at the concentration of maximum boiling point.

(*b*) Minima in dc/dt versus weight fraction curve (fig. 8a, 8b, 9b, 10b) for the mixtures methylene chloride-propionaldehyde, chloroform-acetone, chloroform-propionaldehyde, dichloroethane-ethyl methyl ketone occurs again around the concentration of maximum boiling point. The curve for the mixture tetrachloroethane-Mesityl chloride is flat around the concentration at which boiling point is maximum.

(*c*) In mixtures pentachloroethane-cyclohexanone and chloroform-ethyl methyl ketone, however, no such peculiarity is observed.

These observations show that boiling point and temp. coefficient of velocity have indentical nature with respect to association.

(*ii*) *Legemann's constant*

As mentioned before in chapter III the Legemnan's constant is given by $M^{\frac{1}{2}} \frac{dc}{dt}$. This constant has value around $\frac{-39 \text{ mt gm}^{\frac{1}{2}}}{\text{sec. deg.}}$ for normal liquids, but for associated liquids, it

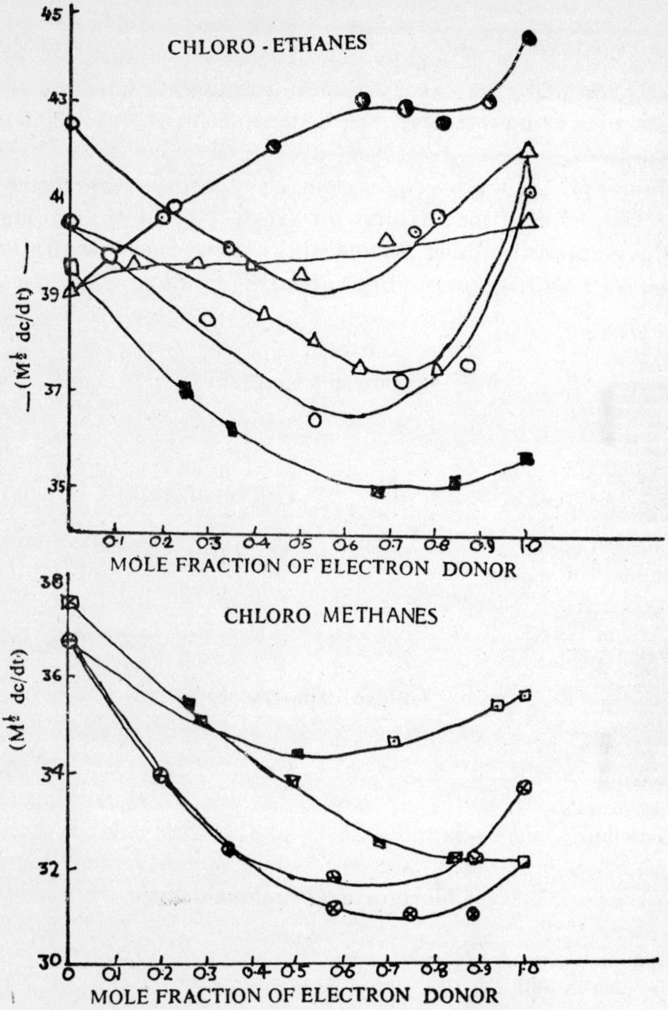

METHYLENE CHLORIDE-PROPIONALDEHYDE ----■
CHLOROFORM - ETHYL METHYL KETONE -------□
CHLOROFORM - PROPIONALDEHYDE ------------⊗
CHLOROFORM - ACETONE---------------------⊗
1,2 DICHLOROETHANE - ETHYL METHYL KETONE--⊠
1,1,2 TRICHLOROETHANE - ETHYL BUTYRATE-----◐
1,1,2 TRICHLOROETHANE - DIOXANE ----------▲
1,1,2,2 TETRACHLORO ETHANE - MESITYL OXIDE---⊙
1,1,2,2 TETRACHLORO ETHANE - CYCLOHEXANONE-△
PENTACHLOROETHANE - CYCLOHEXANONE-------⊙

Fig. 13

attains a value of about 17 to 30 $\frac{\text{mt. gm. } \frac{1}{2}}{\text{Sec. deg.}}$. In the mixture under consideration, as association is present (the components themselves being free of it), the Leigman's constant would show a tendency to decrease (less negative) as the acceptor and donor are mixed together. The Leigman's constant was therefore calculated for all the mixtures for various concentrations. These values appear in table 22 and the curves drawn with these against mole fraction of donor appear in fig 13.

TABLE 22
Legemann's Constant

a. Methyl chloride—propionaldehyde

Concentration of donor in M.F.	0	.26	.40	.68	.85	1.00
Legemann's constant	—37.5	—35.4	—33.8	—32.5	—32.2	—32.1

b. Chloroform—Acetone

Concentration of donor in M.F.	0	.34	.58	.75	.89	1.00
Legemann's constant	—36.7	—32.4	—31.8	—32.3	—32.2	—33.7

c. Chloroform—Propionaldehyde

Concentration of donor in M.F.	0	.2	.34	.58	.75	.89	1.00
Legemann's constant	—36.7	—33.9	—33.4	—31.1	—31.0	—30.9	—32.1

d. Chloroform—Ethyl Methyl Ketone

Concentration of donor in M.F.	0	.29	.52	.71	.86	.93	1.00
Legemann's constant	—36.7	—35.1	—34.4	—34.6	—35.5	—35.4	—35.6

e. 1,2 Dichloroethane—Ethyl Methyl Ketone

Concentration of donor in M.F.	0	.14	.25	.35	.48	.67	.84	1.00
Legemann's constant	—39.5	—37.0	—37.0	—36.2	—35.8	—34.9	—35.1	—35.6

f. 1,1,2 Trichloroethane—Dioxane

Concentration of donor in M.F.	0	.14	.27	.39	.5	.69	1.00
Legemann's constant	—39	—39.6	—39.6	—39.6	—39.4	—40.2	—40.6

g. 1,1,2 trichloroethane—Ethyl Butyrate

Concentration of donor in M.F.	0	.22	.44	.63	.73	.81	.9	1.0
Legemann's constant	—39.0	—40.8	—42.0	—43.1	—43.0	—42.6	—43.1	—44.4

h. 1,1,2,2 tetrachloroethane—cyclohexanone

Concentration of donor in M.F.	0	.42	.53	.63	.68	.72	.80	1.00
Legemann's constant	—40.5	—38.6	—38.1	—37.5	—37.3	—37.1	—37.4	—42.2

i. 1,1,2,2 tetrachloroethane—Mesityloxide

Concentration of donor in M.F	0	.08	.30	.53	.72	.87	1.00
Legemann's constant	—40.5	—39.7	—38.5	—36.3	—37.2	—37.5	—41.2

j. Pentachloroethane—Cyclohexanone

Concentration of donor in M.F.	0	.2	.34	.58	.75	.82	1.00
Legemann's constant	—42.5	—40.6	—40.0	—38,8	—40.4	—40.7	—42.2

It is seen that for the mixtures with acceptor chloroform, dichloroethane, tetrachloroethane, and pentachloroethane, the curve goes through a minima. For those, with acceptor trichloroethane, it undergoes inflexion whereas for the mixture methylene chloride-propionaldehyde, a negative deviation from linearity occurs in the curve. In all the three types of the curves the tendency of the constant is to decrease slightly (less negative). The presence of association is thus implied.

Conclusions (From Sections A and B)

(1) The velocity deviation studies indicate that with the increase in temp. the over all interaction decreases.

(2) The compressibility studies suggest that as the temp. rises, hydrogen bonds get weakened and ultimately get broken, thus making the dipole interaction more predominant.

(3) The temp. coefficient of velocity and the Legemann's constant which are negative for donor and acceptor in their pure state, show a tendency to become less negative in comparison to linear value as the two components are mixed together. This confirms the presence of hydrogen bonds in all the mixtures.

PART III

TEMP. COEFFICIENT OF VELOCITY IN AQUEOUS SOLUTION

Aqueous solution of organic liquids have been of interest to ultrasonic workers[103–119] for the past two decades. The effect of addition of an organic liquid is to break the association in water and when the liquid itself is associated, the two components break each other's association and sometimes these molecules of opposite species indulge in contra association also. Subrahmanian and Raghunath observed while studying the aqueous soln. of alcohols such tendency to break and contra association. Studies of temperature coefficient of velocity were also made by many workers. Giacomin[109] studied water-ethylalcohol mixture and showed that the temp. coefficient of velocity changes gradually with increasing amounts of ethyl alcohol from a positive value for water to a negative value for pure alcohol. Willard[104] while studying the alcohol-water system made a similar report of the behaviour of temp. coefficient of velocity. Nozdrev and Larinov[106] studied temperature coefficient of velocity $\left(\frac{1.}{C}\frac{dc}{dt}\right)$ covering nearly all the members of the homologous series of monohydric alcohols (methyl, ethyl, propyl, butyl) with water cover a full range of concentration and over a broad range of temp. His report was also similar to those of Giacomin and Willard. Now from the studies mentioned in section B of this chapter, we have seen that the temp. coefficient of velocity is a parameter capable of indicating the presence of hydrogen bonds in mixtures. However, none of the above mentioned workers have utilised dc/dt for such studies of aqueous solns. It was therefore, thought worthwhile to use this parameter in the study of formation and breaking of hydrogen bonds in case of aqueous solution.

Referring to the work of Mc Dade[120] et al, velocity in water 0 to 74°C increases reaching a maximum of 74°C. From 74°C to 220°C, there is decrease in velocity in a non-linear fashion but above 220°C the variation is linear with a temp. coefficient of velocity, dc/dt/, of--4.3 mt/sec. deg. Such abnormal behaviour was explained by Mickailov[121] as arising due to competition between two processes viz :

(*i*) Conversion of hydrogen bonded structure to close-packed ones (that is breaking up these bonds) and

(*ii*) increase of mean distance between the molecules with rise of temperative. Between 0° and 74°C, the former predominates where as from 74°C to 220°C, the influence of former vanishes and hence above this temp. water can be assumed to be completely close... packed (There is no hydrogen bond present in water above 220°C).

Thus, in absence of hydrogen bonds in water, the temp. coefficient of velocity for water would have been—4.3 mt/sec. deg. Srinibas Panda[110] et al while studying the effect of dissolving sodium chloride in water-dioxane mixture on velocity versus concentration curve found that the effect of dissolving salt in the aqueous mixture is to reduce the gradient of velocity versus concentration plot for all the concentration below 60% by volume of dioxane. But after 60% concentration (which is 22 mole fraction of dioxane) dissolving of salt has no influence over the velocity of mixture that is velocity versus concentration axis. This led him to conclude that at and beyond 2 mole fraction of non-electrolyte (viz. dioxane) majority of water molecules are in close packed state and water behaves almost like a non-electrolyte. At 9 mole fraction of non-electrolyte, therefore, the conversion of hydrogen bonded structure to close-packed structure can definitely be considered to be complete.

From the above investigations, followings can be expected of the nature of temp. coefficient of velocity for aqueous mixtures : (*i*) At about 2 mole fraction of organic substance, a change in the course of the curve (dc/dt versus concentration) can be expected. (*ii*) Near the region where concentration of water is very low, the mixture would be constituted of close-packed water (with temp. coefficient of —4.3) and the organic substance. In case there is no hydrogen bonding between the two components (that is between the close packed water molecules and the molecule of organic compound) the temp. coefficient of mixture would be given by

$$\left(\frac{dc}{dt}\right)_{\text{mixt}} = 4.3\,x_1 + \left(\frac{dc}{dt}\right)_{\pi nd\ \text{Comp.}} X_2 \qquad \ldots(47)$$

Where X_1, X_2 are the mole fraction of water and organic component and $\left(\frac{dc}{dt}\right)_{\pi nd\ \text{Comp.}}$ is the temp. coefficient of velocity of the organic component. Any divergence of the observed temp. coefficient of velocity of mixture from this value can be explained in light of the formation or breaking of hydrogen bonds.

Such studies were, therefore, undertaken for the mixture of water with methyl alcohol, ethyl alcohol, propyl alcohol, Glycerine, acetone and diethylamine. These mixtures have been studied by a number of workers but only the recent and reliable data of Kukukies[111] and Barfield[112] et al was consi-

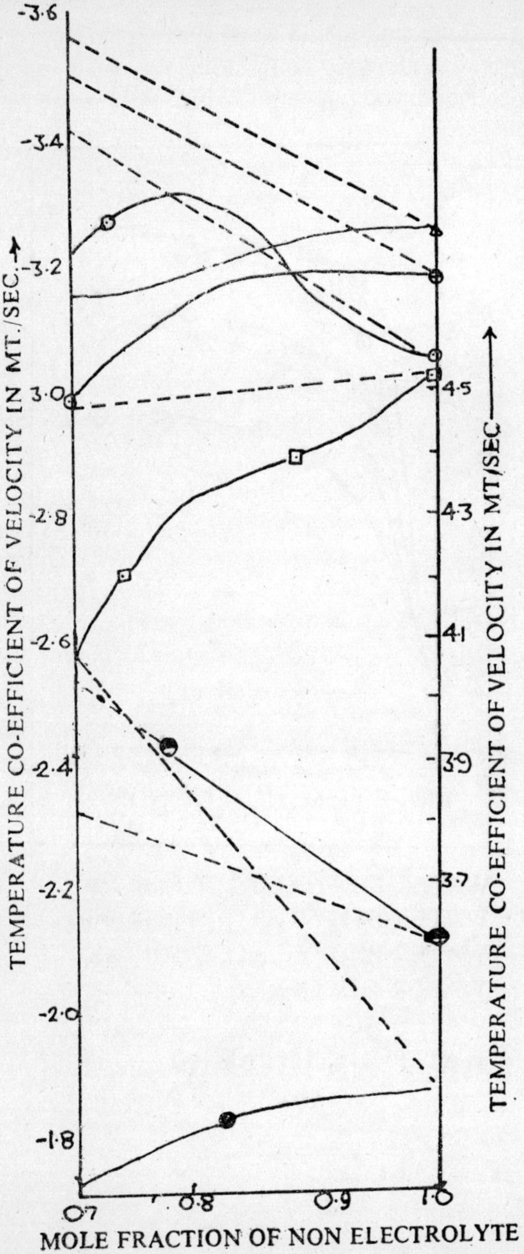
-3·6
-3·4
-3·2
3·0
-2·8
-2·6
-2·4
-2·2
-2·0
-1·8
4·5
4·3
4·1
3·9
3·7
0·7
0·8
0·9
1·0
TEMPERATURE CO-EFFICIENT OF VELOCITY IN MT./SEC.
TEMPERATURE CO-EFFICIENT OF VELOCITY IN MT/SEC
MOLE FRACTION OF NON ELECTROLYTE

(a)

Fig. 14 (a)

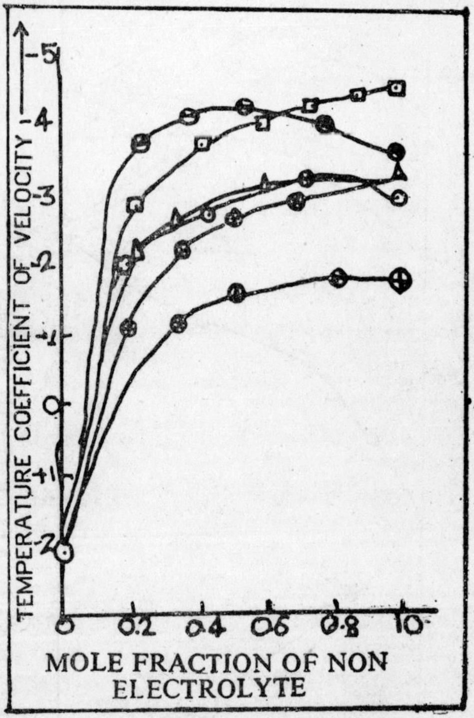

WATER-METHYL ALCOHOL—⊗
WATER-ETHYL ALCOHOL—⊙
WATER-PROPYL ALCOHOL—△
WATER-GLYCERINE—⊕
WATER-ACETONE—⊡
WATER-DIETHYLAMINE—●

(b)

Fig. 14 (b)

dered for velocity at various temperatures. The range of temp. covered was 20°C to 50°C the last one is for the range of 0 to 45°C). From these data, temperature coefficient of velocity at various concentrations was computed. These are plotted against mole fraction of organic liquid in fig. 14 (a) and 14 (b). Later is for the full range of concentration and former (*i.e.* fig. 14 a) is only for the concentration range of 0.7 to 1.0 mole fraction of the organic component. The values of temp. coefficient of velocity for the mixtures also appear in table[23] alongwith the experimental (as read from the graph in fig. 14 a) and theoretical (calculated by using the relation $\left(\frac{dc}{dt}\right)$ mixt.) values of temperature coefficient of velocity for the concentration of 9 mole fraction of organic component for all the mixtures.

Results and discussion

From the study of fig. 14 b, one finds that in all the aqueous mixtures, the temp. coefficient of velocity versus mole fraction curve steeply changes from a positive value to a negative value till the concentration of 0.2 mole fraction of organic component is reached. After this concentration, the curve takes a new course, becoming flat and ultimately reaching 1 mole fraction value of the organic component in different ways depending upon the component. This is exactly what has been predicted by the author. It appears from the figure that upto 0.2 mole fraction of the second component (*i.e.* the organic component) the concentration of ice like structure to close packed structure of water is very rapid. Beyond 0.2 concentration, majority of the water molecules are in close packed state (that is free of hydrogen bonds).

Now concentrating upon the second component rich region of the mixtures (fig. 14 a), it can be seen that from 0.9 to 1.0 mole fraction of the organic substance, temp. coefficient of velocity versus concentration curve is almost linear.

For concentration less than 0.9 mole fractions curve goes through a maxima in case of the mixtures water-ethyl alcohol, water-diethylamine. This must be due to the fact that the temp. coefficients of velocity for the two organic components (−3.0

TABLE 23

Temp. coefficient in Aqueous solutions of Non-Electrolytes

Non-electrolyte	*Concentration of non-electrolyte (mole fraction)*	*temp. coefficient of velocity*	*temp. coefficient of velocity at .9 M.F. of the organic component*	
			Experi-mental	*Theore-tical*
Diethylamine	0	+2.2		
	0.24	—3.7		
	0.37	—4.1		
	0.55	—4.2	—3.7	—3.6
	0.78	—4.0		
	1.0	—3.6		
Acetone	0	+2.2		
	0.22	—2.9		
	0.42	—3.7		
	0.60	—4.0	—4.4	—4.5
	0.74	—4.2		
	0.88	—4.39		
	1.00	—4.52		
Methyl alcohol	0	+2.2		
	0.22	—2.1		
	0.34	—2.6	—3.1	—3.2
	0.6	—3.1		
	1.0	—3.26		
Ethyl alcohol	0	+2.2		
	0.19	—2.0		
	0.42	—2.7	—3.1	—3.1
	0.73	—3.36		
	1.0	—3.04		
Propyl alcohol	0	+2.2		
	0.2	—1.1		
	0.36	—2.1		
	0.5	—2.6	—3.2	—3.3
	0.7	—2.98		
	1.0	—3.18		
Glycerine	0	+2.2		
	0.36	—1.1		
	0.52	—1.6	—1.8	—2.1
	0.82	—1.82		
	1.0	—1.86		

and 3-6 respectively) are less negative than the value —4.3 for close-packed water molecules.

From a comparison of experimental and theoretical values of dc/dt at 9 mole fraction of the second component (Table 23), one finds that :

(*i*) In water-ethyl alcohol mixture, the two slopes are equal. Hence there occurs no contra-association between the two components.

(*ii*) Similarly in water-diethylamine, the experimental slope is even more negative than the theoretical one. This again shows the absence of contra-association.

(*iii*) In glycerine-water mixture, the experimental slope (—1.8) is much more than the theoretical one (—2.1), and hence the presence of hydrogen bonds between glycerine and water molecule, is implied because if there had not been any such bond, the curve would have shown even a maxima because temp. coefficient of velocity for glycerine (which is —1.8) is very much more than —4.3 (the value for unhydrogen bonded water molecule), there should be an increase in the negative value of dc/dt as a little water is added to glycerine. The contra-associations seems to be quite strong so as to suppress any initial increase in negativity of dc/dt of the mixture arising due to the above mentioned reason.

(*iv*) For the same reason mentioned above, a maxima in dc/dt versus concentration plot is expected in mixtures : water-methyl acohol, water-propyl alcohol (since dc/dt for both the alcohols is less negative than —4.3 mt/sec. deg. and also both are associated) but dc/dt values are found to be less negative than —4.3 for both the solutions. Hence in these cases, a little contra-association seems to be present.

(*v*) Acetone is an unassociated liquid with dc/dt quite close to the value —4.3 mt/sec deg. Hence no maxima in the curve can be expected. In case, the contra-association is strong, the experimental value of dc/dt at .9 concentration would be very much less negative than the theoretical one. But this is not found to be the case. The former is found to be only a little less negative than the later. Hence in this mixture of water-acetone, contra-association seems to be very weak.

These conclusions arrived at from the temp. coefficient of velocity studies are quite in accordance with the nature of the organic components, e.g. taking the two extreme cases, (viz. diethylamine and glycerine). Diethylamine is an unassociated liquid. Hence it cannot form strong hydrogen bonds with the close-packed water molecules whereas in glycerine there are three 'OH' groups which readily associated with water molecules making the experimental temp. coefficient less negative than the theoretical one.

Conclusion

The prediction of breaking and formation of hydrogen bonds in a mixture from temp. coefficient of velocity studies seems quite satisfactory. Thus, temp. coefficient of velocity determination can replace other methods in study of association.

4

Ultrasonic Study of Phases in Solutions

IT HAS been observed in general that near every transition from one state to other, ultrasonic parameters exhibit peculiar behaviour followed by a discontinuity at the transition point in the curve of the particular parameter against the temp. For instance, the Russian Scientists[44b] observed in abnormal behaviour of the ultrasonic velocity near the solidification point in liquids such as benzene, carbon tetrachloride, ethyl alcohol, carbon bi-sulphide, nitrobenzene and halides of benzene. It was found that the velocity versus temp. curve becomes non-linear as the solidification point is approached. Measurement of velocity at low temp. down to $-70°C$ were carried out by them for a number of saturated hydro-carbons, acetates, formates, and carbon tetrachloride and again it was observed that whenever the temp. went down to as far as the solidification point, the divergence from linearity exceeded the experimental error. Schaafs[122] and Woelk[14] observed a jump of velocity at the normal boiling point and melting point for many normal liquids. Another study that was made was near the liquid to crystalline state by Koshikin and coworkers[128] who noted an abrupt increase of velocity and absorption in benzene, o p-xylene, cyclohexane, thymol, p-dichlorobenzene, biphenyl etc. In the supercooled region of liquids, Legemann[124] et al, Broni[125], et al, Litovitz and Parthasastry[48b] et al observed a similar break followed by a change of slope in velocity versus temp. curve on either side of the melting point.

In study of the critical temp. region, by Schneidev, Breazeale, Nosdrew, Schaafs, Tenneberger, it was observed that the velocity versus temp. curve exhibits a very sharp minima at critical temp. of the medium critical region of mixtures were also studied.[126–128] The mixtures studied were Methyl alcohol-hexane, Methyl alcohol-heptane, aniline-hexane, triethylamine-water, Methanol-cyclohexanol.

The transition that is of current interest to ultrasonic workers namely change of mesophase to isotropic phase in a liquid crystal also showed unusual behaviour. Gabrielli[42], et al observed in the liquid crystal of p-amexyanisol (having the transition temp. of 135°C) that the velocity versus temp. curve has an unsymmetric temp. at 135°C. In the liquid crystal of cholesteryl benzoate, the nature of the velocity versus temp. was found to be straight line on either side of the transition point having different slopes. Padmini,[49(ii)] et al observed a similar type of discontinuity in the curve for cholesteryl formate. However, for cholosteryl acetate, they observed a minima in vicinity of liquid crystalline phase transition. Nature of the curve was found to depend upon the type of liquid crystal (which are of three types viz : Cholesteric, Sonectic, nematic). Recently Natal[49(iii)] et al studied nematic crystals. The order of the decrease in velocity near the transition point was found to be quite large in all the liquid crystals (about 100 mt/sec) unlike other transitions. These abnormalities of velocity and allied parameters near the transition point were attributed by Frenkel[129] to the prominent presence of pre-transitional effects at transition temp. He showed theoretically that as the transition temp. is reached, the groups of molecules of second phase exist in first phase.

Around the liquid glass transition temp. also, velocity showed a marked behaviour. While studying the liquid glass transition temp by using ultrasonic technique, it was found that by properly choosing the period of ultrasonic waves shorter than the relaxation time associated with structural changes, the glass like properties could be observed even at temp. much higher than the normal glass transition temperature. Litovitz while studying such pseudo glass transition for glycerol and arochlor found that the velocity versus temperature curve suffers a change of slope at the new pseudo transition temperature.

It is thus clear from these studies (that is of melting point, boiling point, supercooled region, critical temp region, around the transition from mesophase to isotropic phase in liquid crystals, liquid-glass transition temperatures) that accurate measurements of ultrasonic propagation constants provide an excellent means of studying the transition between the regions where a change of phase occurs.

No work appears to have been done to study the behaviour of a solution, when the solute changes its phase from any solid to liquid (as a result of rise in temp. of the solution through the melting point of the solute). A review of ultrasonic work indicates that although the mechanism of a solute going into a solution is the same irrespective of it being a liquid or solid, still these two solutions behave differently. For example, ultrasonic velocity and Rao's constant at infinite concentration of solid state in solvent are independent of the nature of solvent and are nearly same as those for the solute had it existed as a liquid at the temp. under consideration. This indicates that there occurs a little interaction in a solution of solid solutes. The same should be true of a solution, the solute of which is in liquid state. But there are many instances available where the velocity versus concentration curve exhibits deviations from linearity even in a dilute solute. In such cases, the velocity at 100 per cent concentration of solute obtained by extrapolation of such a curve will not give the velocity of solute at the concerned temp. Thus a solution, having its solute in solid state differs from that whose solute is in liquid state. Hence choosing low melting solutes and making velocity study at temperatures on either side of the melting point of the solute, the exact difference in the behaviour of solid solute and liquid solute and the type of transition between the two phases can

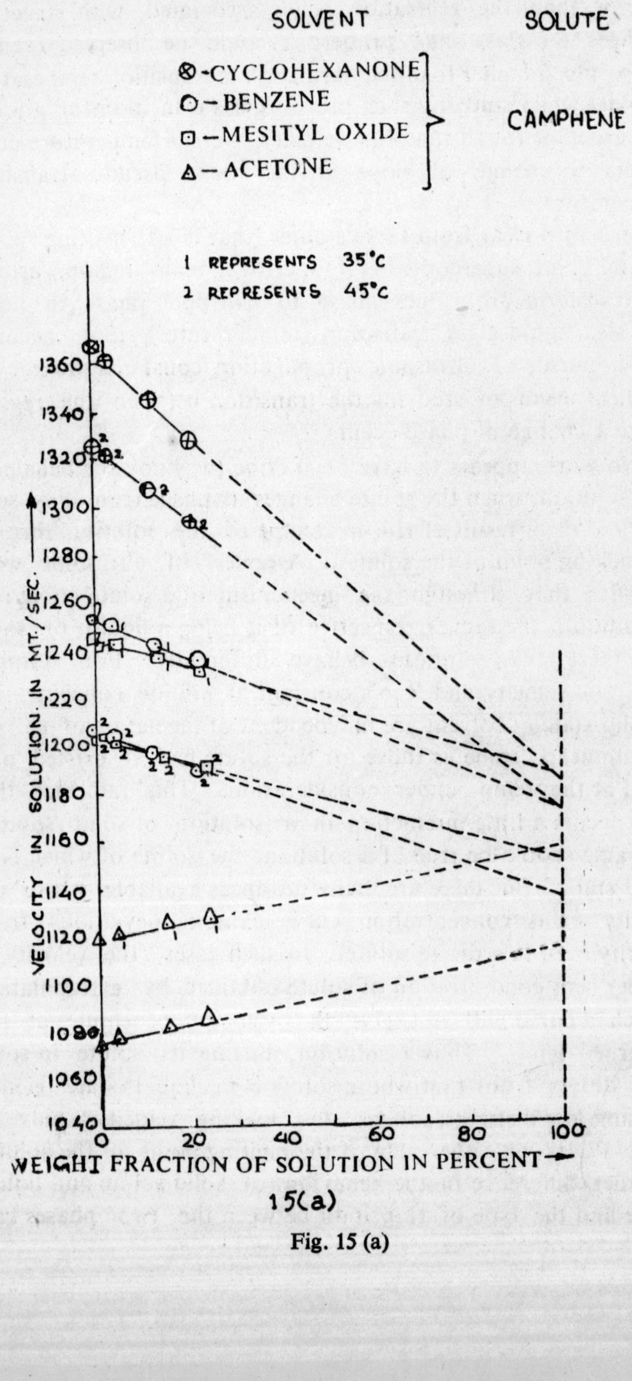
SOLVENT
SOLUTE
⊗-CYCLOHEXANONE
O-BENZENE
□-MESITYL OXIDE
Δ-ACETONE
CAMPHENE
1 REPRESENTS 35°C
2 REPRESENTS 45°C
VELOCITY IN SOLUTION IN MT./SEC.
WEIGHT FRACTION OF SOLUTION IN PERCENT
15(a)

Fig. 15 (a)

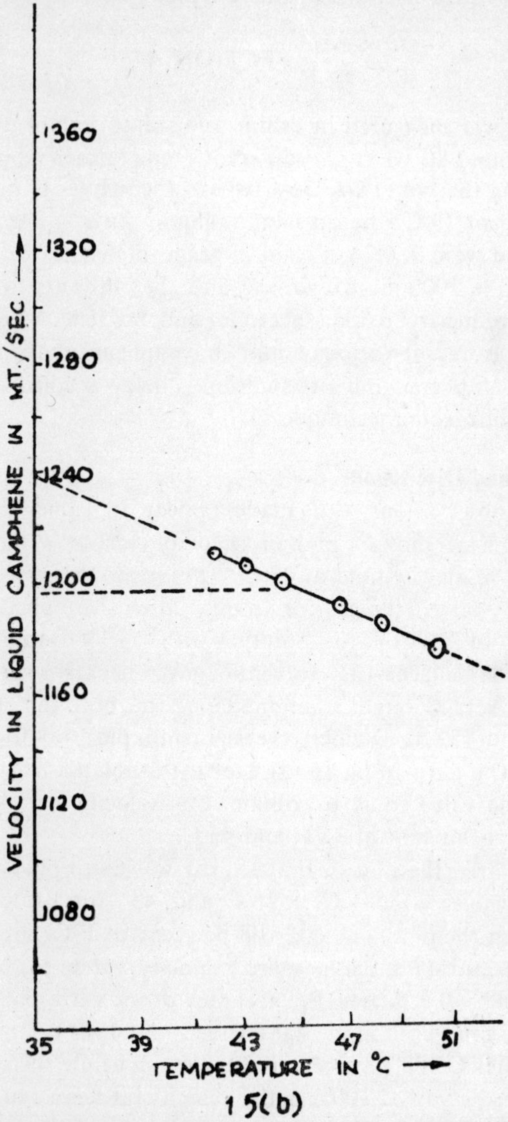
1360
1320
1280
1240
1200
1160
1120
1080
VELOCITY IN LIQUID CAMPHENE IN MT./SEC →
35
39
43
47
51
TEMPERATURE IN °C →
15(b)

Fig. 15 (b)

be observed. Such studies were, therefore, undertaken in the following three sections of this chapter.

SECTION A

Velocity was measured in soln. of solute camphene (having M.P. around 40°C) at two different temperatures (concentration remaining the same) viz. 35°C when the solute is in its solid state and at 45°C when it is in its liquid state. The cancentration used were 0, 0.25, 1 and 2 gram molecules of camphene dissolved in 1000 ml of solvent and the solvents were cyclohexanone, mesityl oxide, acetone and benzene. Velocity was also measured at various temp. in camphene above its melting point. Camphene and its solution is being studied for the first time by ultrasonic technique.

Results and Discussions

The observations, thus made, appear in Tables 24 and 25. Figure 15 (a) shows a plot of velocity versus weight fraction of solution at 35°C and at 45°C. Figure 15 (b) gives a plot of velocity versus temp. for pure liquid camphene. Unlike majority of solution, the solution with cyclohexanone, mesityl oxide and benzene as solvents have negative gradient for velocity versus weight fraction curve at both the temp. (i.e. 35° C and 45°C). Velocity versus temp. plot was found to be linear. The plots in fig 15 (a) were extra-polated to 100 per cent camphene value so as to obtain the velocity (c 100 per cent) in pure camphene at 35°C and 45°C.

Similarly, the plot in Fig. 15 (b) was extra-polated to 35°C and the values velocity C_t at 35°C and 45° for camphene were read from the plot. These C 103 per cent and C_t are presented in Table 26 taking the measured velocity values to be accurate to within ± 0.4 mt/sec. the accuracy of the extra-polated values C 100% comes out as ± 4.0 mt/sec. Analysis of the Table 26 shows that C 100% values are very much below their respective C_t values. Now C 100% is the velocity (at temp.) of camphene when it is in solution (the concentration of solvent being zero) and C_t is the velocity of pure camphene at temp. t. Since C 100% are less than C_t values, interaction between solute and

TABLE 24

Velocity in solutions of camphene for various concentrations at 35°C and 45°C

solution	*Concentration of solute in percentage weight fraction*	*Temp.* 35C°	45°C
Camphene in Cyclohexanone	0	1366.9 mt/sec	1326 mt/sec
	3.5	1361	1321
	12.0	1345.6	1307
	21	1328	1294
Camphene in benzene	0	1353.2	1206.5
	3.7	1250	1204 mt/sec
	13	1242	1197.6
	23	1236	1190
Camphene in Mesityl oxide	0	1245	1204
	3.4	1242	1201
	14	1237.5	1196.0
	22	1232	1192
Camphene in Acetone	0	1119.6	1175.1
	4	1121	1178
	14	1124.8	1081.6
	25	1128	1086

TABLE 25

Velocity in pure camphenat various temps. above its melting point

temp. °C	41.8	43	44.4	46.6	48.2	50.5
Velocity in mt/sec	1211	1206	1200	1192	1186	1176

solvent appears to be present, whatever be the state of solute. Atthough the plots at two temperatures for variours solutions appear to be similar to each other and seem to throw no light on the nature of the state of solute, a close scrutiny of these, however, reveals that they differ in following respects : for any particular solution, the velocity versus concentration curve has more slope at 35°C (that is, when the solute is in solid state), then at 45°C (when the state of solute is liquid) whatever be the gradient of slope whether positive or negative. This makes the velocity versus concentration plot at 35°C for all the solutions more convergent than that at 45°C, with the result that the extrapolated values C 100% are more close to each other when the solute is in solid state than when it is in liquid state. This shows that in a solution, the interaction present is less dependent upon the nature of solvent when solute is in solid state than when the solute is in its liquid state.

TABLE 26

Velocity C 100% and C_t for camphene

Solution	C *100%*		*Average* C *100%*		Ct	
	35°C	*45°C*	*35°C*	*45°C*	*35°C*	*45°C*
Camphene in Cyclohexanone	1184	1173				
Camphene in benzene	1173	1135	1175	1143	1239	1198
Camphene in Mosityl oxide	1188	1152				
Camphene in Acetone	1156	1114				

SECTION B

The work mentioned in the above section concerning the behaviour of solution of camphene in various solvents is only at two temperatures. In the present section the results of study of a number of solutions at various temp. below and above the M.P. of solute has been discussed. In this work, the

SOLUTION OF DEPHENYLAMINE IN ACETONE

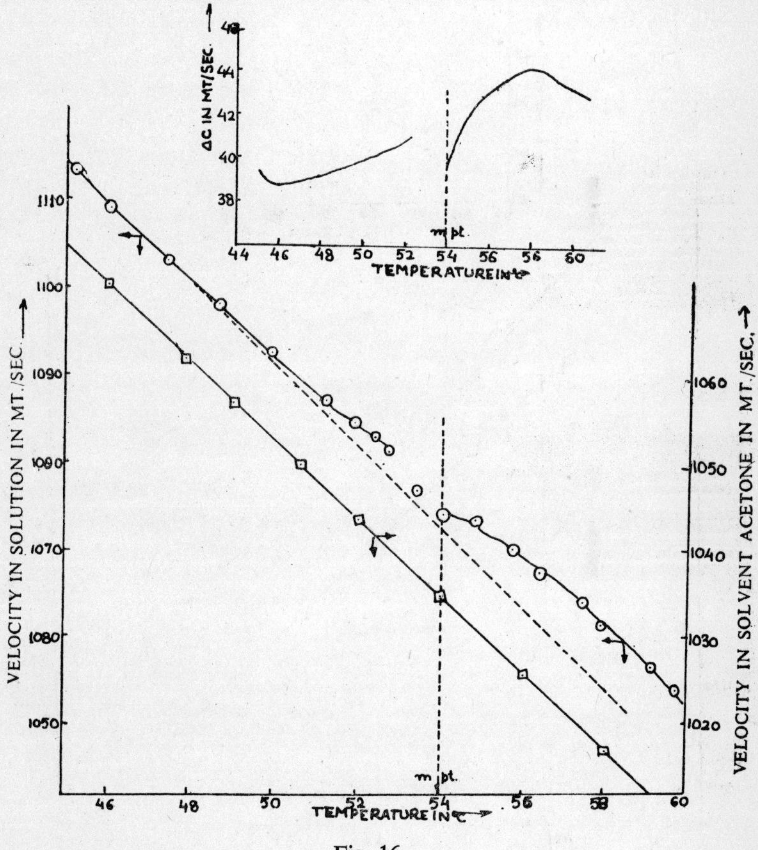

Fig. 16

SOLUTION OF DIPHENYLAMINE IN BENZENE

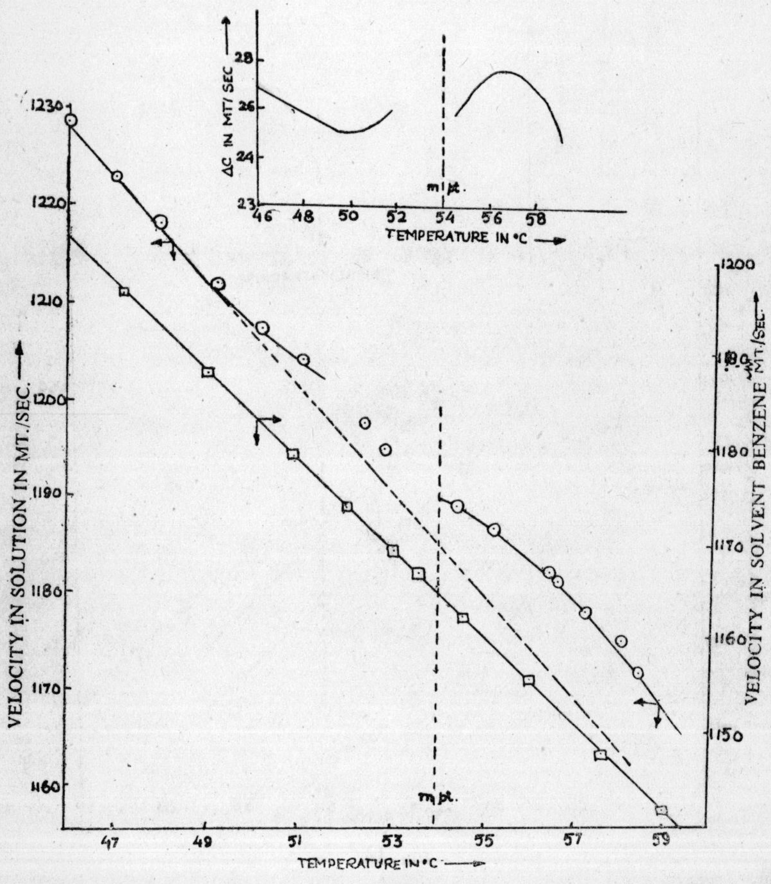

Fig. 17

SOLUTION OF DIPHENYLAMINE IN CHLOROFORM

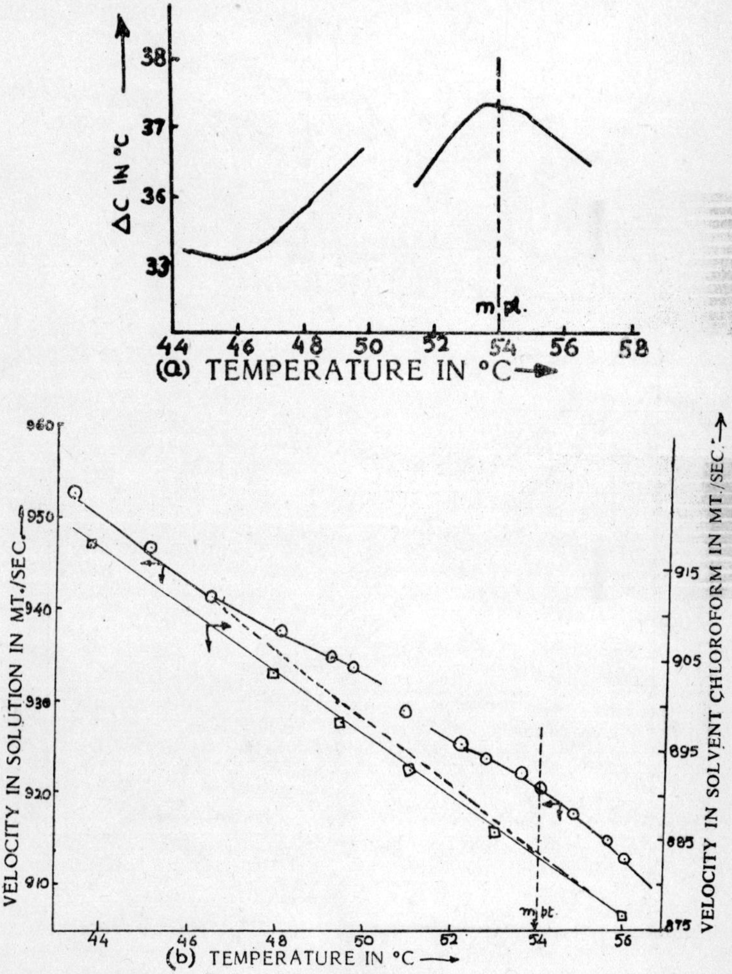

Fig. 18

SOLUTION OF BENZOPH NOCE IN ACKTONE

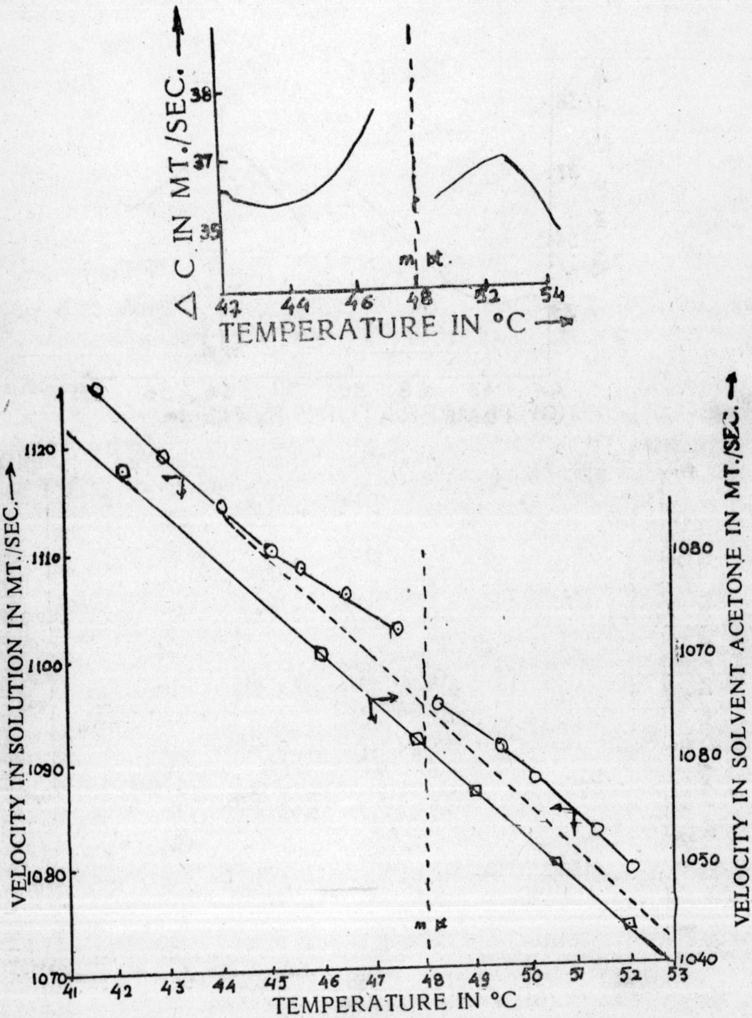

Fig. 19

SOLUTION OF BENLOPHENONE IN BEHZENE

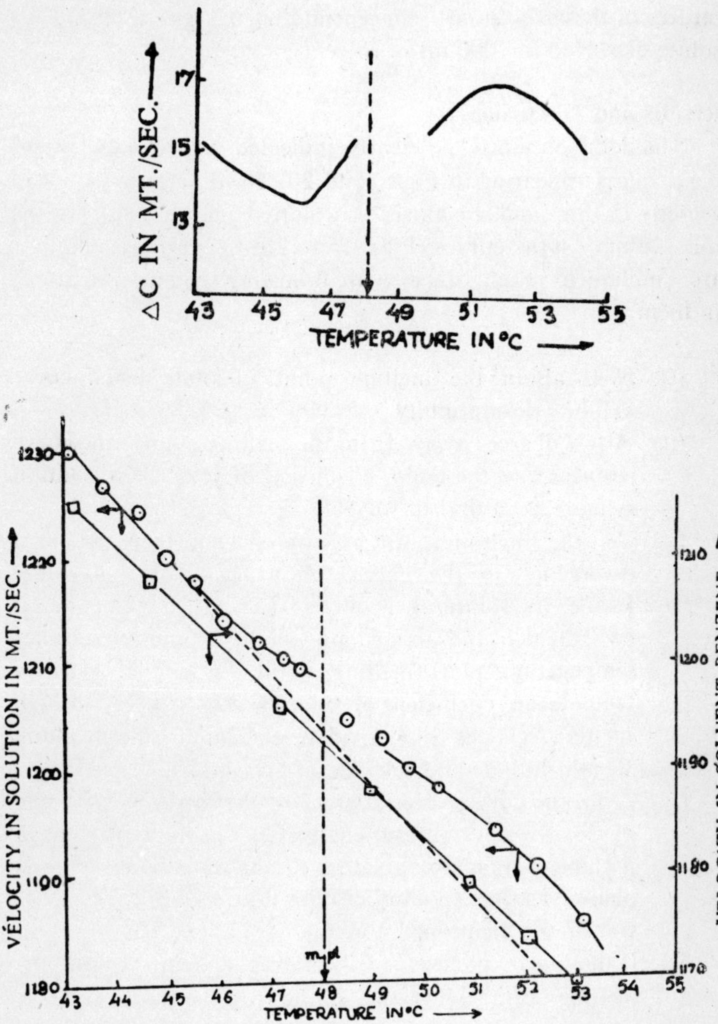

Fig. 20

difference in behaviour of solid and liquid solute is more clear. The solute chosen were : diphenyl amine (M.P. 54°C) benzo phenone (M.P. 48°C) and the solvents were benzene, acetone, chloroform ultrasonic velocity was measured at various temperatures in these solutions (concentration : 0.5 gm molecule of solute dissolved in 1000 ml of solvent).

Results and Discussions

The data obtained is clearly indicated in velocity versus temp. plots appearing in Figs. 16 to 20. The difference ($\triangle c$) in velocity in soln. and solvent is also plotted against temp. in the some plot. All the curves (Figs. 16 to 20) for various solutions are similar to each other with following common features in them :

(*i*) Near about the melting point of solute, each curve exhibits discontinuity.

(*ii*) A few degrees away from the melting point, the curve is linear and the temp. coefficient of velocity in solution is more than that in solvent.

(*iii*) As the melting point is approached, there occurs, a divergence in the curve. As a result of this solute (while in solution) acquires a positive temperature coefficient in the immediate vicinity of the ($\triangle c$ versus temperature plot) melting point. A similar positive temperature coefficient of velocity was observed in high melting glasses like oxides of Silicon, Germenium, Boron, Sulfide and selenide of Arsenic.

(*iv*) From the curves (figs. 16 to 20), it is obvious that the phase of a solution also changes as the solute present in it changes its phase. Nature of the transition from one phase of solute to another does not resemble any of the transitions mentioned above.

(*v*) If the linear portion of the curve at temp. below the melting point (when the solute is in its solid state) is extrapolated to any temp. above the M.P. of the solute this extra-polated line is found to lie below curve in the liquid solute region. This implies that velocity at any temp. in a soln. is more when the solute is in liquid state than that had the same solute existed as solid at that temp. That is, the interaction in a solution is more when the solute is in liquid state than when it is in solid state.

The feature (ii) of the curve must be due to the fact that with rise in temp., solvent expands. This causes the concentration of solute to decrease and hence there is an additional decrease in velocity of solution (than when the solvent is free of solute). This makes the dc/dt of solution more than that of solvent.

SECTION C

The work mentioned in this section has been undertaken with a view to see if the lowering of melting point of solute due to the presence of impurity in it is apparent while in solution. The substance camphene has been chosen as solute because of its low M.P. (42.7°C) and high value of molecular depression constant, K = 350°.

According to Raoult's law, when W gms of any impurity (having M. wt. equal to M) is added to W gms of comphene, there occurs a depression in the value of melting point of camphene and is given by

$$\triangle T = \frac{100\,K}{m}\left(\frac{W\text{-}}{M}\right) \qquad \text{...(48)}$$

In the present work, W has been taken as 0.5109 gm (0.25 gm molecule of camphene dissolved in 1000 ml of benzene) W has been so chosen that the depression $\triangle T$ may be 13°C and 8°C (this value of W was calculated by using the above relation) the impurities chosen were p— toludene diphenyl amine, benzophenone. 0.5109 gm of camphene (for $\triangle T$ = 0 the amount of camphene taken was 2.0436 gms. in 15 ml of benzene) was mixed with adequate amount of impurity and gently heated till the mixture melted. The melt was shaken well and cooled. The solid solution so formed was then used as solute and it solution made in 15 ml. of benzene. Velocity

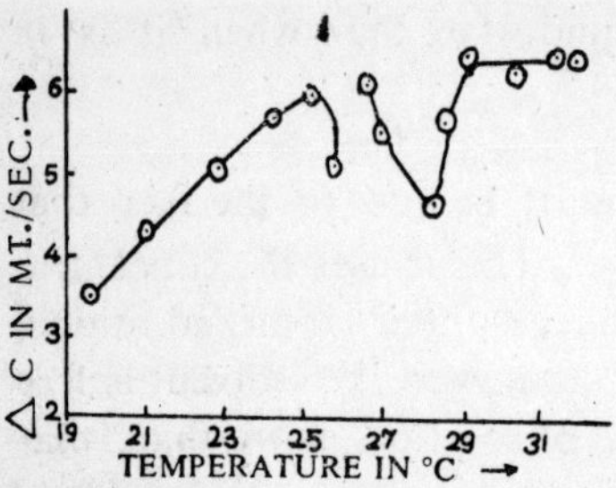

(a) SOLUTION OF CAMPHENE-P. TOLUIDINE IN BENZENE FOR △T=13°C

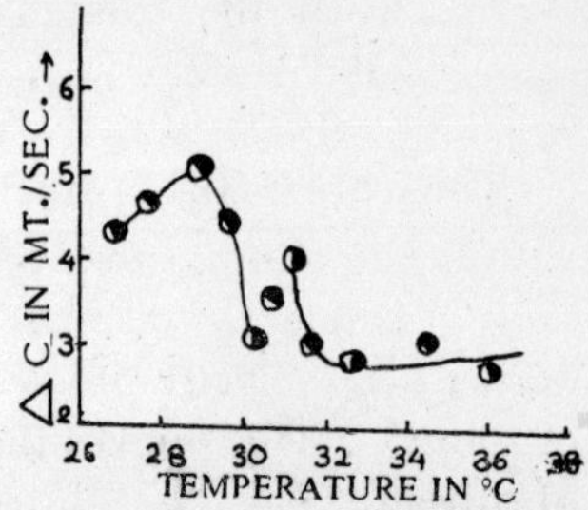

(b) SOLUTION OF CAMPHENE-P. TOLUIDINE IN BENZENE FOR △T=8°C

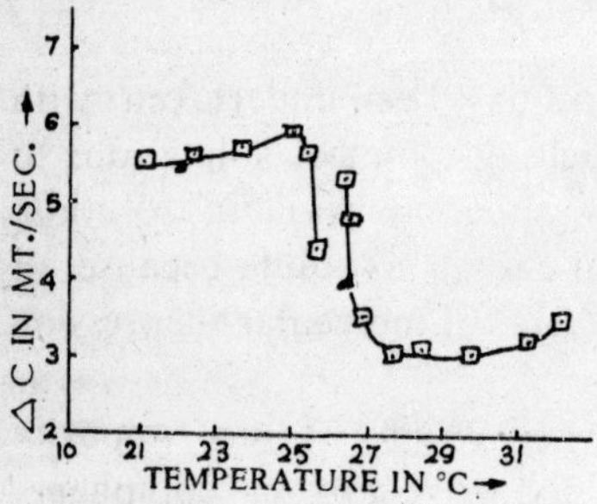

(c) SOLUTION OF CAMPHENE-DIPHENYLAMINE IN BENZENE FOR △T=13°C

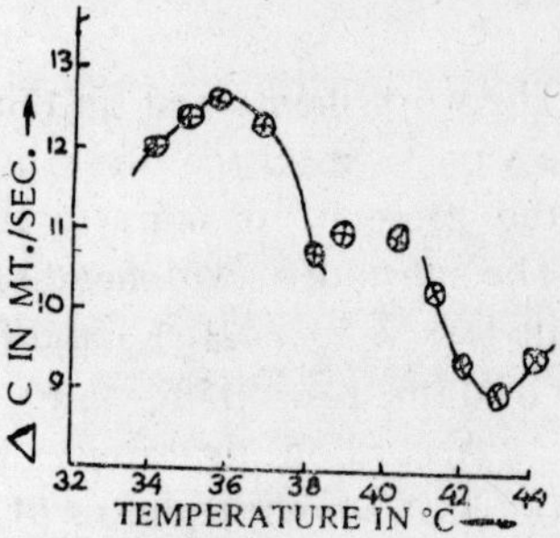

(e) SOLUTION OF COMPHENE IN BENZENE FOR △T=13°C

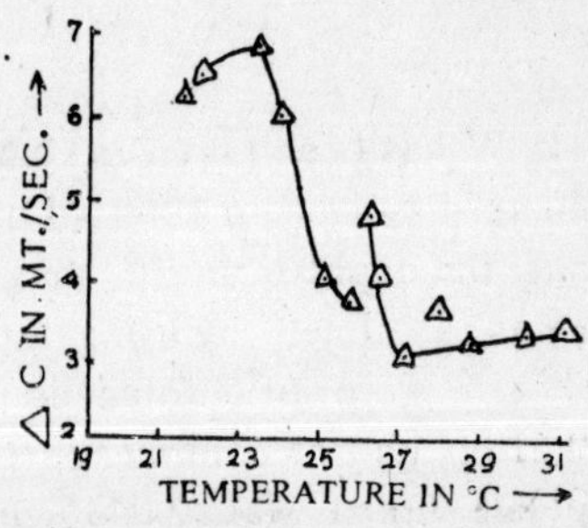

(d) SOLUTION OF CAMPHENE-BENZOPHENONE IN BENZENE FOR △T=°C

Fig. 21

was measured around the temperature where the pseudo melting point was likely to be observed.

Results and Discussions

The data obtained in solvent for various temperatures is indicated in Figures 21(a) to 21(e) plots of $\triangle c$ against temperature. Nature of the curve $\triangle c$ versus temperature is found to be just the reverse of what it is for solutions mentioned in section B (this is because the velocity in camphene is lower than that in benzene at any particular temp).

In case of the solution of camphene in benzene for $\triangle T = 0°$ [Fig. 21(e)] the break in the curve occurs at 38°C. Taking this as the melting point of camphene for the depressions $\triangle T = 13°$C and for 8°C, the pseudo melting point should be 25°C and 30°C respectively. Now referring to the curves in fig. 21 a, c, d, it can be seen that the break in curve (for $\triangle T =$ 13°C) occurs at 26°C for all the three impurities. This value is quite close to the theoretically calculated value of 25°C. Similarly the break in the curve (for $\triangle T = 8°$C) fig. 21b occurs at 31°C, which is again nearly equal to the theoretically alculated value of 30°C.

Although the break around the melting point is only 1 mt./sec. still as the experiment was carried out at lower temperature, and also as the maxima in the micrometer were quite sharp for these solutions, the accuracy of velocity measurement could be taken as $\pm$ .2 mt/sec.

Thus, ultrasonic technique helps in locating the normal and Pseudo melting points.

Conclusion

From the work presented in three sections of this chapter, following conclusions can be drawn :

(1) In a solution, interactions are present among its components whatever be the state of solute. This interaction, when the state of solute is solid, is less dependent on the nature of solvent than it is when the state of solute is liquid.

(2) The interaction at a particular temp. in a solution is more when solute is in liquid state than when it is in solid state

TABLE 27

The Difference in Velocity (mt/sec) in solution and in solvent benzene at various temperatures

a. Impurity : P—toluidine = 13°C

Temp., °C	19.6	21.0	22.8	24.2	25.2	25.8	26.6	27.0	28.2	28.6	29.2	30.4	31.4	32.0
	3.5	4.3	5.1	5.7	6.0	5.1	6.2	5.5	4.6	5.6	6.5	6.3	6.5	6.5

b. Impurity P — toluidine = 8°C

Temp., °C	26.8	27.6	28.9	29.6	30.3	30.6	31.1	31.6	32.6	34.4	36.0
	4.3	4.6	5.0	4.4	3	3.5	4	3	2.8	3	2.7

c. Impurity Diphenylenine = 13°C

Temp., °C	20.8	22.2	23.4	24.8	25.2	25.4	26.2	26.3	26.6	27.4	28.3	29.6	31	32
	5.5	5.6	5.7	5.9	5.7	5.4	5.3	4.8	3.5	3.1	3.1	3.1	3.2	3.5

d. Impurity — Benzophenone = 13°C

Temp., °C	21.4	21.8	23.2	23.8	25	25.6	26	26.2	27	27.8	28.6	30	31
	6.2	6.5	6.8	6.0	4	3.7	4.8	4.0	3.0	3.6	3.2	3.3	3.4

e. Impurity — Nil = 0°C

Temp., °C	34	34.8	35.6	36.8	38.2	39.0	40.4	41.4	42.2	43.2	44
	12.0	12.4	12.6	12.3	10.7	11	11	10.3	9.4	9.0	9.5

(as observed by extrapolation of velocity versus temp. curve in solid solute region to liquid solute region).

(3) Solutes, while in solution, retain their identity as indicated by the breaks at the respective melting points of solutes. A solution changes its phase as the solute changes its phase from solid to liquid.

(4) Even the lowering of melting point due to presence of impurity in solute is manifested when the solute is in solution. This shows that a solute continues to act as solvent to impurities present in it even after entering into solution in another solvent.

(5) The transition accompanying the change of state of solute in a solution does not resemble any of the transitions mentioned in the introduction part of this chapter. The behaviour of velocity near the melting point should be due to the group of molecules of second phase existing in the first phase. A thorough study of many more solutions with much more precision instrument for velocity (and hence allied parameters) measurement is likely to help in understanding the exact nature of interaction taking place during the transition.

(6) Break in the velocity versus temp. curve does not occur exactly at the melting point of a solute. It is likely that the behaviour near the transition point is a function of time, as observed in the pseudo glass transition studies. By choosing the time scale of measurement (that is by choosing the period of ultrasonic waves shorter or longer than the relaxation time), it may be possible to observe shifting of melting point of solute in the solution. This opens a new field for absorption studies.

(7) From the data presented in sections B and C, it appears that the drop of velocity at the M.P. is proportional to the M.P. of solute, e.g. when the melting point is around 26 to 31°C, the drop is around 1 mt/sec, whereas for solution having melting point of solute 48 to 54°C the drop is 3-4 mt/sec. More precise measurement of velocity and more data is, however, needed to corroborate this statement.

5

Ultrasonic Velocity and Boiling Point

SOME workers have suggested an identical behaviour of compressibility, freezing point and boiling point, when their variation with certain parameter is studied. For example Weister[130] has shown for chlorides of elements of fourth groups that the compressibility and the freezing point curves when plotted against molecular weight show exactly identical variation. Again Gabriell et al, from the study of various mixtures have suggested that the compressibility versus concentration plots, and boiling point versus concentration plots for binary mixtures, may show similarities. Lagemann and Dunber have suggested an empirical relation between the Rao's constant and boiling point as :

$$\text{Log}\, R = A + BT_h \qquad \ldots(46)$$

The validity of this relation has been tested for some of the homologous series.

No work however seems to have been done yet, to obtain theoretically a correlation between the boiling point and the sound velocity. An attempt, therefore, has been made in this chapter to obtain such a theoretical relation and to test its validity. The relation obtained is based on the Eyring' rate theory and the hole structure of liquids.[131] The former theory gives a relation between viscosity and boiling point and the later one connects fluidity with free volume.

Starting from Telang's[132] absolute viscosity equation (which is based on ths Eyring's rate theory), Palit[133] deduced the following equation correlating the viscosity η the surface tension r.

$$\log_{10}(\eta M) = K_1 V^1 (MV_{sp})^{\frac{2}{3}} + K_2$$

where K_1, K_2 are constants whose values are given by

$$K_1 = 1.091\ N^{\frac{1}{3}}/2.303\ RT = 0.165\times 10^{-2}\ (\text{at } 20^\circ C)$$

$$K_2 = \log_{10}(hNhSP)\ (V_{sp} - h_{sp})^{\frac{4}{3}}$$

where *hsp* is the Vander Waal's constant per gram. The available experimental data were found to be in very good agreement with equation 47. Combining this equation for Log_{10} with Eotvos equation.

$$\sqrt{}^1\ (MVsp)^{\frac{3}{2}} = K_E\ (T_c - T)$$

and Guldberg's equation $T_c = \frac{3}{2} T_h$

where T_h is the boiling point (at 760 mm) in °k. Palit arrived at the following expression for a homologous series :

$$\log_{10}(\eta M) = m_1 T_h + m_2$$

where $m_1 = \frac{3}{2} K_1 K_E = 5.2\times 10^{-3}/\text{dig}\ (\text{at } 20^\circ C)$

$$KE = 2.1$$

$$m_2 = (K_2 - K_1 K_E T)$$

and

Thus, according to above relation, m_1 should be 5.2×10^{-3}/deg at 20°C for all the homologous series if the following assumptions are correct; (i) Eotvos and Gulberg's equation are valid, (ii) KE has same value for all the members. Such conditions are satisfied approximately by unassociated liquids but not by associated ones. Palit demonstrated the linearity of log (mm) versus Tb curve for C_{17}–C_{35} hydrocarbons at 20°C and the slope from experiment came out as 6.7×10^{-3}.

Now coming to the relation between free volume and fluidity, the relation obtained by Macleod[134] is based on 'Eyring's hole theory' but for two modifications (i) The viscous flow of liquid (which was neglected by Eyring in his relation) has been taken into consideration (ii) Eyring assumed that the effect of increase of temp, is to increase the free volume by introduction of holes which have definite size compared to that of the molecules. Macleod, on the other hand, has considered free volume as being continuous so that the differential equation may be used freely in all the calculations.

With these modifications, Macleod showed that the fluidity ϕ is related to free volume Vf by ϕ = Abf. Where A is a constant of proportionality. Asoke Kumar Mukerji[135] showed that η for a number of liquids is nearly unity and he further suggested that η will differ from unity in case there is any change in molecular weight due to association. Thus, the final relation between fluidity and free length takes the form:

$$\phi = \frac{1}{\eta} = AV_f \qquad ...(54)$$

Derivation of the relation between the Sound Velocity and the Boiling point

Since the free volume Vf is related to sound velocity by

$$C = \left(\frac{V}{V_f}\right)^{\frac{1}{3}} \left(\frac{RTV}{M}\right)^{\frac{1}{2}}$$

$$V_f = \frac{V}{C^3} \left(\frac{RTV}{M}\right)^{\frac{3}{2}} \qquad ...(55)$$

From equation for ϕ

$$-\log \eta = \log A + \log V_f \qquad ...(56)$$

and also

$$\log_{f^v} = \log\left(\frac{V}{C3}\right)\left(\frac{V}{M}\right)^{\frac{3}{2}} + \tfrac{3}{2} \log RT$$

on substitution one gets,

$$-\log \eta = \log A + \log \left(\frac{V}{C^3} \frac{V}{M}\right)^{\frac{3}{2}} + \tfrac{3}{2} \log RT$$

$$\text{or } \log \eta m = -\log A - \log \frac{V}{C^3} \frac{V^{\frac{3}{2}}}{M^{\frac{5}{2}}} - \tfrac{3}{2} \log RT$$

Substituting this in equation

$$-\log_{10} A - \log_{10} \frac{V}{C^3} \frac{V^{\frac{3}{2}}}{M^{\frac{5}{2}}} - \tfrac{3}{2} \log_{10} RT = m_1 T_h + m_2$$

$$\text{or, } \log_{10} \frac{C^3 M^{\frac{5}{2}} \rho}{M V^{\frac{3}{2}}} = m_1 T_b + m_2 + \log_{10} A + \tfrac{3}{2} \log RT$$

$$\text{or, } \log_{10} \frac{C M^{\frac{1}{2}} \rho^{\frac{1}{3}}}{V^{\frac{1}{2}}}$$

$$= m_1 T_b + m_2 + \log_{10} A + \tfrac{3}{2} \log RT$$

$$\text{or, } \log \frac{C M^{\frac{1}{2}} \rho^{\frac{1}{3}}}{V^{\frac{1}{2}}} = \frac{m_1}{3} T_b + \left(\frac{m_2}{3} + \frac{\log_{10}}{3} + \tfrac{1}{2} \log_{10} RT \right)$$

$$\text{writing, } \left(\frac{m_2}{3} + \frac{\log_{10} A}{3} + \tfrac{1}{2} \log_{10} RT \right) = B$$

(B would be a constant for members of a homologous series at a particular temperature), we get finally the relation between the velocity C and the boiling point Tb as

$$\log_{10} \left(\frac{C M^{\frac{1}{2}} \rho^{\frac{1}{3}}}{V^{\frac{1}{2}}} \right) = \frac{m_1}{3} T_b + B \qquad \text{...(57)}$$

A plot of log 10 $(CM^{\frac{1}{2}}P^{\frac{1}{3}}/V^{\frac{1}{2}})$ versus boiling point Tb should therefore be a straight line. The slope should be one third of that of the slope of log (ηM) versus Tb plot (Ref. equation 50) for all the homologous series of nornal liquids. Since $m_1 = 5.2 \times 10^{-2}$ per degree at 20°C the expected slope is 1.7×10^{-3} per degree. In order to test the validity of the relation, velocity, density and boiling point data was collected for a number of homologous series. As the value of ratio of specific heat in not readily available and further as it does not contribute much to the left hand side of the above equation log 10 $(V^{\frac{1}{2}})$ has been neglected and the following equation has been used :

$$\log_{10} (CM^{\frac{1}{2}}P^{\frac{1}{3}}) = \left(\frac{m_1}{3} T_b + B \right) \qquad \text{...(59)}$$

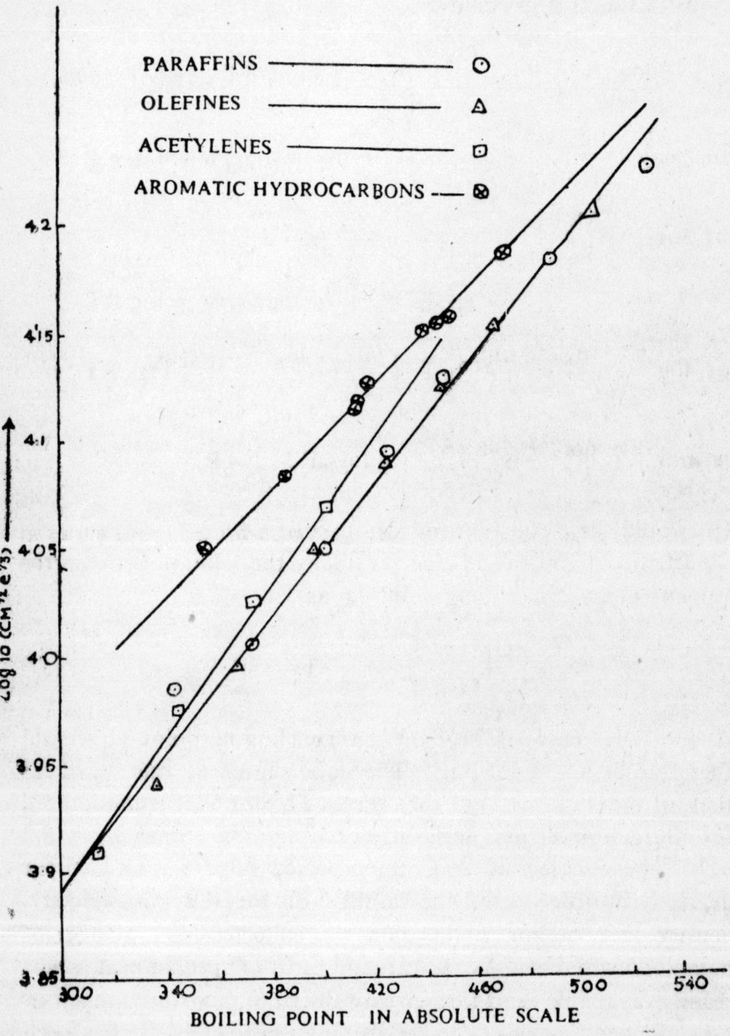
PARAFFINS
OLEFINES
ACETYLENES
AROMATIC HYDROCARBONS
4.2
4.15
4.1
4.05
4.0
3.95
3.9
3.85
300
340
380
420
460
500
540
BOILING POINT IN ABSOLUTE SCALE

Fig. 22

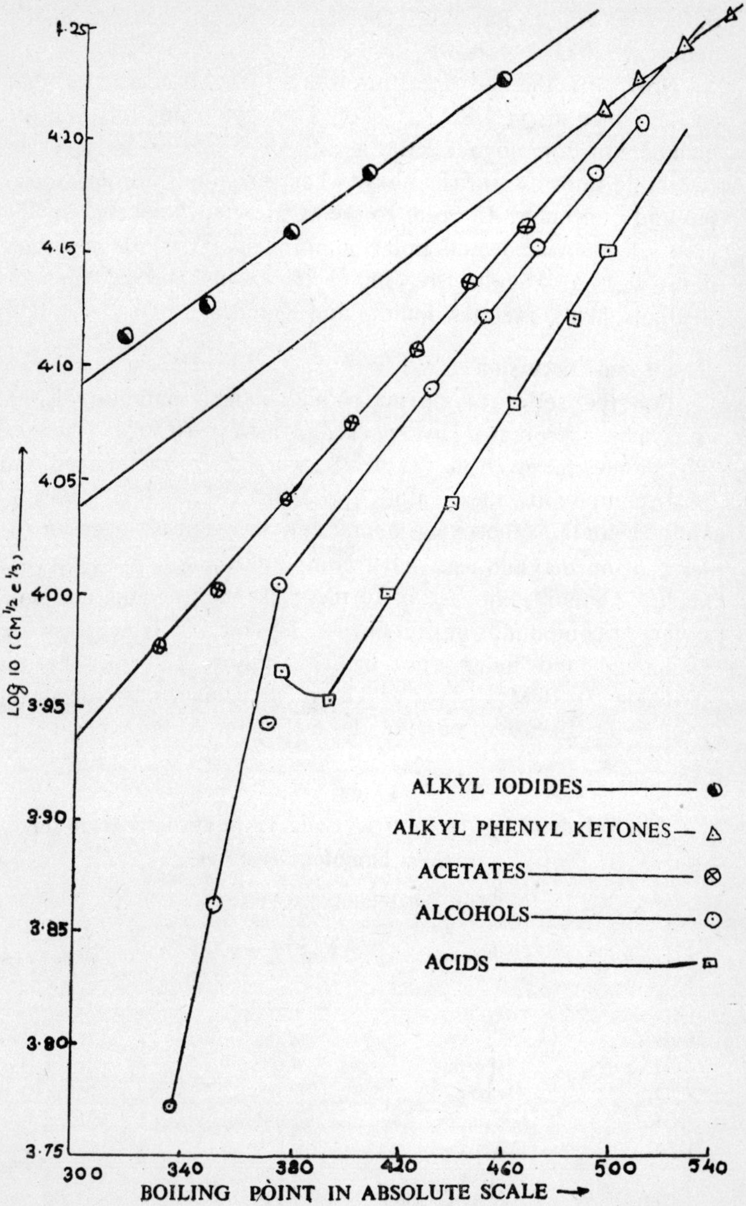
4.25
4.20
4.15
4.10
4.05
4.00
3.95
3.90
3.85
3.80
3.75
LOG 10 (CM$^{1/2}$ e$^{1/3}$) →
300
340
380
420
460
500
540
BOILING POINT IN ABSOLUTE SCALE →
ALKYL IODIDES
ALKYL PHENYL KETONES
ACETATES
ALCOHOLS
ACIDS

Fig. 23

Validity of this equation was tested for the members of the following series: Paraffin, Olefin, Acetylene and Aromatic hydrocarbons.

Now, for the above relation to hold true, it is essential that (i) KE (since $m_1 = \frac{3}{2}$ K, KE) should be the same for all the members of homologous series and (ii) η should be unity. These two conditions are satisfied only when the liquids are unassociated and non-polar, In order to see as to what happens in the case of associated and polar homologous series, the necessary data was also collected for the following series: Aliphatic alcohols, acids, acetates, iodides and Ketones.

Result and Discussion

For the series of normal liquids namely paraffin, Olefine, Acetylene, Aromatic hydrocarbon, the log 10 ($CM^{\frac{1}{2}}P^{\frac{1}{3}}$) and Tb values are given in Table 28 whereas for associated and polar compounds, these values appear in Table 30. Tables 30 and 31 contain the experimental and theoretical slopes for the series of normal and associated compounds. Log 10 ($CM^{\frac{1}{2}}P^{\frac{1}{3}}$) versus Tb curve in case of former class (i.e.) normal and non-polar) of compounds appears in fig. 22 and for later class (i.e. associated and polar one) fig. 23. Figure 22 depict that the plots log 10 ($CM^{\frac{1}{2}}P^{\frac{1}{3}}$) versus Tb are all straight lines for all the four series. In case of paraffin and olefine series, the experimental

TABLE 28

The values of log 10 (C $\rho^{\frac{1}{3}}{}_{m}{}^{\frac{1}{2}}$) and Tb in absolute scale for various homologous series

a. Paraffine T = 20°C

Formula	*Bompound*	*Log 10 AP $M^{\frac{1}{2}}$*	*Boiling point °K*
C_5H_{12}	Pentane	3.873	309
C_6H_{14}	Hexane	3.984	341.7
C_7H_{16}	Heptone	4.005	371
C_8H_{18}	Octane	4.055	399
C_9H_{20}	n-Nonane	4.096	423
$C_{10}H_{22}$	n-Decane	4.129	446
$C_{12}H_{26}$	n-Dedecane	4.187	487
$C_{14}H_{30}$	n-tetradecane	4·233	525.5

b. Olefines T = 20°C

Formula	*Compound*	*Log 10*	*Boiling point*
C_6H_{12}	Hexane	3.940	336.6
C_7H_{14}	Heptene	3.996	366.3
C_8H_{16}	Octen	4.050	395
C_9H_{18}	Nowen	4.090	421.9
$C_{10}H_{20}$	Decen	4.127	445
$C_{11}H_{22}$	Undecen	4.157	561-465
$C_{13}H_{26}$	Tridecen	4.209	505.7

c. Acetylenes T = 20°C

Formula	*Compound*	*Log 10*	*Boiling point*
C_5H_6	Pentin	3.908	313
C_6H_{10}	Hexin	3.975	344
C_7H_{12}	Heptin	4.027	372
C_8H_{14}	Octin	4.067	400

d. Arometic hydrocarbon T = 20°C

Formula	*Compound*	*Log 10*	*Boiling point*
C_6H_6	Benzene	4.049	353
$C_6H_5CH_3$	Toluene	4.084	384
$C_6H_4(CH_3)_2$	O-Xylene	4.128	416
	M-	4.119	412
	P-	4.116	411
$C_8H_4(CH_3)_3$	Mesitylene	4.152	437
	Pseudocumal	4.156	443
	Hemillithol	4.159	448
$C_6H_4(CH_3)_4$	tetramethyl benzene	4.188	468

TABLE 29

The value of slope M_1 at 20°C

Series	*Experimental M_1*	*Theoretical M_1*
Paraffin	4.2×10^{-3} per deg	
Olefine	4.5×10^{-3} per deg	5.2×10^{-3} per deg.
Acetylene	5.4×10^{-3} per deg	
Aromatic hydrocarbon	3.6×10^{-3} per deg	

b. Alcohols T = 20°C

Formula	*Compound*	*log 10*	*Boiling point*
CH_3OH	Methanol	3.769	338
C_2H_5OH	Ethanol	3.863	351
C_3H_7OH	Propanol	3.945	370
C_4H_9OH	n-Butanol	4.007	373
$C_6H_{13}OH$	n-Hexanol	4.096	428
$C_7H_{15}OH$	n-Heptanol	4.131	448
$C_8H_{17}OH$	n-Octanol	4.162	467
$C_9H_{19}OH$	n-Nonanol	4.195	478
$C_{10}H_{21}OH$	n-Decanol	4.128	505

c. Acid T = 20°C

Formula	*Compound*	*log 10*	*Boiling point*
$H\ COOH$	Formic aeid	3.969	374
$CH_3\ COOH$	Acetic acid	3.956	392
$C_2H_5\ COOH$	Propionic acid	4.003	414
$C_3H_7\ COOH$	Butyric acid	4.046	437
$C_4H_9\ COOH$	Valeric acid	4.090	459
$C_5H_{11}\ COOH$	Caproic acid	4.128	481
$C_6H_{13}\ COOH$	Heptoic acid	4.163	492
$C_7H_{15}\ COOH$	Capryl acid	4.189	512

TABLE 30

Values of log 10 ($C\rho^{\frac{1}{3}}{m}^{\frac{1}{2}}$) and boiling point in absolute scalc for various polar and associated Homologous series

a. Alkyliodides T = 20°C

Formula	*Compound*	*log 10*	*Boiling point*
CH_3I	Methyliodide	4.116	315
C_2H_5I	Ethyliodide	4.131	345
C_3H_7I	Propyliodide	4.163	375
C_4H_9I	n-Butyliodide	4.191	403
$C_6H_{13}I$	n-Hexyliodide	4.235	453

b. Alkyl Phenyl Ketone T = 30°C

Formula	*Compound*	*log 10*	*Boiling point*
$C_6H_5COC_2H_5$	Ethyl Phenyl Ketone	4.225	491
$C_6H_5COC_3H_7$	Propyl Phenyl Ketone	4.241	504
$C_6H_5COC_4H_9$	Betyl Phenyl Ketone	4.256	521
$C_6H_5COC_5H_{12}$	Amyl Phenyl Ketone	4.272	538

c. Acetates T 30°C

Formula	*Compound*	*log 10*	*Boilig npoint*
CH_3COOCH_3	Methylacetate	3.978	330
$CH_3COOC_2H_5$	Ethylacetate	4.003	350
$CH_3COOC_3H_7$	n-Propylacetate	4.044	374
$CH_3COOC_4H_9$	n-Butylacetate	4.080	399
$CH_3COOC_5H_{11}$	n-Amylacetate	4.114	422
$CH_3COOC_6H_{13}$	n-Hexylacetate	4.144	442
$CH_3COOC_7H_{15}$	n-Heptylacetate	4.171	464

TABLE 31

The values of slope M_1

Series	*Temperature*	*Experimental M_1*	*Theoretical M_1*
Alkyl Iodide	20°C	2.7 per deg	5.2 per deg
Alkyl Phenyl Ketone	30°C	2.82	3.9
Acetates	30°C	4.2	3.9
Alcohols	20°C	19.2, 4.8	5.2
Acids	20°C	5.7	5.2

points fall over the same straight line. Now regarding the slope, the experimental and theoretical slopes of log 10 $(CM^{\frac{1}{2}}P^{\frac{1}{3}})$ versus Tb curve, are quite close to each other for paraffins, Olefine and Acetylene.

For Aromatic hydrocarbon series, the experimental slope is somewhat less than the theoretical one. Thus, for homologous series of normal compounds, the relation seems to hold true.

For associated and polar compounds, however, it can be seen from the fig. 23, that for series of polar (unassociated) compounds, the experimental points fall on straight lines. For iodide and ketone, the experimental slope is quite less than the theoretical value. For acetate also, the agreement is good. In case of alcohol and acid series, the curve exhibits abnormality. For alcohols, the curve consists of two straight lines with different slopes upto butyl-alcohol the curve has a slope unusually larger than the theoretical one. After butyl alcohol it bends and acquires a slope quite close to the theoretical value. Thus, the relation seems to hold true, butyl alcohol onwards. Again, in acids, for formic acid and acetic acid, the curve has a negative gradient (i.e. a minima occurs) but acetic acid onwards, it has a linear nature with slope, again close to the theoretical value.

Conclusions

(1) The relation derived above for velocity and boiling point is satisfied by normal liquids of a homologous series like paraffins, olefines, etc,

(2) The relation seems to be applicable for even certain polar unassociated series of compounds like acetates.

(3) However, in case of associated homologous series such as alcohols and acids, the relation seems to be valid for higher members only.

REFERENCES

1 Ch. Kittal J. Chem. Phys. 14, 1946, 614.
2. L. Tonks Phys. Rev. 50 ; 1936, 955.
3. B. Jacobson Acta Chemical Scand 6, 1952, 1485.
4. H. Eyring and J. F. Koncaid J. Chem. Phys 6, 1938, 620.
5. Yasuka Wada J. Phys. Soc. Japan 9, 1954, 641
6. J. Frenkel, Kin theory of liquids Oxford Univ. Press 1946 P. 174 fl.
7. J.S. Rowlinson and C.F. Curtiss J. Chem. Phys. II, 1951, 1519.
8. L. Hall, Phys. Rev. 73, 1951, 1519.
9. W. Schaafs Z. Phys. 114, 1939, 251.
10. Arakawa, Kiyoshi, Kojema, Kazumitsu Bull. Chem. Soc. Japan 48(i), 1978, 26-8
11. (a) S. Prathasarathy, Proc. Ind. Acad. Sci.

(i) 2, 1935, 497
(ii) 3, 1936, 519
(iii) 4, 1937, 59
(b) S. Parathasarathy, Current Science 6, 1938, 322

12. E. Froyer, J.C. Hubbard, D. Andrews J. Am. Chem. Soc. 51, 1929, 759
13. R.T. Lagemann, D.R. Mc. Millan, W.E. Woolf J. Chem. Phys. N.Y. 17, 1949, 369
14. H.U. Woelk Dissertation T.U. Berlin 1961
15. H.U. Woelk Dissertation T.U. Berlin 1962.
16. R. Kuhnkies and W. Schaaffs, Acustica

(i) 13, 1963, 407
(ii) 16, (1965/66), 66

17. R.T. Lagemann Woolf, Evan Underwood J. Am. Chem. Soc. 70, No. 9, 1948, 2995
R.T. Lagemann, D.R. Mc Millan, M. Woolsey J. Chem. Phys. 16, 1948, 247.
18. (i) R.T. Lagemann, J. Am. Chem. Soc. 73, 1951, 5891
(ii) R.T. Lagemann. J. Am. Chem. Soc. 73, 1951, 3213

19. R.T. Lagemann J. Am Chem. Soc. 74, 1952, 1602
20, G.W. Marks J. Ac. Soc. Am. 41, 1967, 103
21. R. Prashad Ind. J. Phys. 15, 1941, 323, 19, 1945, 47
22. W. Schaaffs Ergch exakt Natur W. 25, 1951, 110
23. R. Nulsch—Kuhnkics, Acustica 15, 1965, 383
24. Thamsen Joern, Acta, Chem, Scand, 19(8), 1965, 1939-50
25. Thamsen Joern, Acta, Chem. 195, 1950, 145-69 Kolloid—z.4, 1951 122.

26. (i) V.F. Nozdrew and A.F. Lependin Zhur. Fiz, Khim 27 (No. 8) 1953, 1257
(ii) B.B. Kudriyavtsev. Akusi. Zhur 2 (No. 4), 1956, 331
(iii) O. Singh, N. Dass, N.C. Varshneya J. Phys. A5(3), 1972, 37-8

26. (b) S. Parathasarathy and N.N. Bakhshi Proc. Soc. B. Vol. LXVI, 1953, 368

27. R.T. Legemann and W.S. Dubber J. Phys. Chem. 49, 1945, 428

28. A. Weissler J. Am. Chem. Soc. 70, 1948, 1634
z h z d 71, 1949, 93

29. A. Weissler J. Am. Chem. Soc. 71, 1949, 419

30. R.T. Legemann. J. Am. Chem. Soc. 70, No. 9, 1948, 2996

31. P.R.K.L. Padmini, Trans. Faradoy Soc. 56, 1960, 1404-8

32. Enzo-Butta, Ricerca Sci 26, 1956, 3643-7

33. A. Weissler. J. Am. Chem. Soc. 71, 1949, 1272

34. M.I. Salakhutdinov Dokl. Akad. Nank Tadzh S.S.R. 11(2), 1968, 22-5

35. Rama Rao, Current Science 23, 1954, 325

36. A. Weissler J. Chem. Phys. 15, 210, 1947

37. (i) B.B. Kudriyav Sev Zhur. Fiz, Khim, 28, 1954, 932-5
(ii) Padmini and Ram Chandra Rao Ind. J. Phys. 35, No. 7, 1961, 346
(iii) G.H. Fondeegg. et al Trans. Faraday. Sec. 63(4), 1967, 870-8

38. M. Baccaredda, chimmica e. industria 32, 1950, 155-60

39. V.I. Ryitov, G.S. Yakoleva, L.F. Slonovskii T.I. Golomox. Fiz. Khim Povakh Yavlenii Rasplavakh 1978, 51-5

40. Satish Chandra Shrivastava Ind. J. Phys. 33, No. 11, 1959, 503-4

41. L.G. Melkonyan, Premenenie ultrakustik K. Issiedovan Veshchestva No. 3, 1956, PP. 31-49.

42. S.V. Meniloy Dokl. Akad. Nauk SSSR 181(6), 1968, 1427-9.

43. A.V. Narasemham, B.B. Manikian Ind. J. Phys. 43 (12), 1975, 1068-74

44. (a) Alfred Weissler J.A.C.S. 71, 1949, 1272.

44. (b) O.Z. Giolik, Ukr, Fiz. Zh 7. 1962, 806-12.

45. C. Rao, Raghupathi Rao, P.S.R. Krishna, Current Science 41(8), 1975, 286-8.

46. Kurt Attenberg Z. Physic, Chem. 195, 1950, 145-59.

47. N.F. Otpushchennikov Iso Vysshikh uchen bn. Zavedenii Fiz 1964 (3), 108-16

48. (a) Hadamard Dissertatsiya Thesis 1954.
48. (b) S. Parthasarathy and V.N. Bindal Ind. J. Phys. 34, 1960, 272
49. (i) I. Gabrielli, L. Verdini, Neovo Comento, 4, 1955, 526
(ii) C.G. Kartha and P.R.K.L. Padmini Ind. J. App. Phys. 9, 1971, 723-28
(iii) G.G. Natal, D.E. Commin, Phys. Rev. Letter 28 (22), 1979, 1439-41
50. Frank T. Giikor J.R. J.A.C.S. 55, No. 7, 1933, 2709
51. F.T. Giicker Chem. Rev. 13, 1933, 111.
52. P. Dehye and E. Huckel, 2, Physik, 24, 185, 1923.
53. C.H. Bachem Z.F. Physik 101, 1936, 541
54. B.H. Krishnamurthy Rao J.S.I.R. 98, 1950, 215
55. Scot. Ohenhaus and Wilson, J. Phys. Chem. 38, 1934, 937.
56. K.S. Rao and B.R. Rao J.S.I.R. 178, 1958, 444
57. F. Giicker and R. Haag. J.A.S. Am. 25, 470, 1933.
58. (i) A Moesveld and H. Hardon Z. Physik Chem. A 155, 1931, 238
(ii) A Passynskii Acta Physicochin (URSS) 8, 1938, 385
(iii) A Giacomini and H. Pesce Ricerca Sci. 11, 1940, 605
(iv) Y. Wada and S. Shimba and O. Oda J.A.S.A. 22, 1950, 850.
(v) T. Yasunaga and Tisasaki, J. Chem. Soc. Japan 72, 1951. 87-89
59. (i) P. Padmini and B. Rao Ind. J. Phys. 34, 1960, 565
(ii) G. Marks. J. Ac. Soc. Am. 31, 1959, 936
32, 1960, 327.
60. (i) Mikhailov Proc. 4th international congress on accoustics Copenhagen paper J 43
(ii) P. Phare Trans. Faraday Soc. 58, 1962, b, 363
61. F.H. Stillinger J. Chem. Phys. 35, 1961, 1581
62. T. Veeraih Current Science 35 23, 1966, 589-90
63. (i) S.V. Subramanyam Phys. letters 1, 1962, 90
(ii) Allam and Lee J. Chem. Soc. Supyl. 2, 1969, 6049
(iii) K. Ramabrahman and K. Suryanarayan Ind. J. app. Phys. 6 No. 8, 1968, 422-25
64. S. Lakshmi Shobby, S. Prakash and S. Prakash Z. Physik Chem. 5-6, 1968, 247-52
65. a. A.V. Satyavati, Acustica 19, No. 6, 1967-68, 350.
65. b. W. Schaaffs Z. Physik, 105, 1937, 658
65. c. L. Sibaiya and L.R. Narasemhaya J. Mysore Univ. 1, 1941, 133

65. d. P. Silaramaswamy J.P.A.P. 5(2), 1969, 57-60
66. Rao D. Sc. Thesis Andhra Univ. 1962
67. Lal Ind. J. Phys. 24, 1950, 461
68. K. Rao and B. Rao Proc. Phys. Soc. London 73, 1959, 239
69. S. Rao, Nand, Rao Subha Ind. J.P.A. Phys 10(4), 1972, 314
70. Haxinm Shiio J. Am. Chem. Soc. 80, 1958, 70
71. Haxime Shiio, 77, 1955, 4980
72. Yutaka Miyahere J. Chem. Soc. Japan 73, 1952, 265
73. O. Ncmote J. Phys. Soc. Japan 11, 1956, 1146-52
74. Vasaku Wada Oyo Butsuri 23, 1954, 15-18
75. Kowatska W. Kovalaki, M. Bodzel, G. Kameh Z. Fiir Physikalisch chemi 234 (5/6) 1967, 419
76. B.B Kudriyavtaav and S.Z. Balyan Premenenie ultra akustiki R. Issledovan Veshchestva No. 6, 1958, 79-80
77. a. Yasumage, Bull Chem. Soc. Japan 37 (ii) 1964, 1955-8
77. b. H. Sackmann and A. Bockzok, Z. Physic. Chem. Nane, Folge 29, 1961, 329
78. M.V. Kaulgud, Acustica 10, 1960, 316
79. M.V. Kaulgud Ind. J. Phys. 15, 1941, 323
80. Ram Parshad Ind. J. Phys. 15, 1941, 323
81. Ram Parshad Ind. J. Phys. 16, 1942, 1
82. Ram Parshad Ind. J. Phys. 16, 1942, 307-315
83. M.V. Kaulgud Zeit, Flir, Physikahsche chemic 36 (5-6) 1963, 365.
84. O. Ncmoto J. Phys. Soc. Japan 8, 1953, 553
85. O. Nomoto J. Phys. Soc. Japan 8, 1953, 553
86. (i) Ferdinands Dmesso Atti Accad. Nazl Lincei Rend Classi Sci. Fis mate nat 17, 1954, 109-13
(ii) Ibid ibid 13, 1952, 131-8
(iii) Ibid ibid 10, 1951, 235-42.
(iv) I.G. Polalski and Z.I. Khodov, Akad. Nank ukr S.S.R. No. 5. 1954, 33-44
(v) P.R. Maidu, V.R. Krishnan J. Phys. Sec. Japan 20 (9), 1965, 1554-8
(vi) K.C. Reddy, Trans Faraday Soc. Japan 58, 1962, 2352-7
(vii) K.C. Reddy, S.V. Subrahmanyam. Ball clas sci, Acad. Roy Belg (5) 49 (7), 1960, 230-40
(viii) J. Gabrielli and G. Pocani Pubbl. Fac. Sci e ing Univ. Trieste Ser. B. No. 108, 1952, 3-9
87. (i) K. Subha Rao and B. Ramchandra Rao J.S.I.R. 18 B, 1959, 223-6

(ii) I. Gabrielli and G. Paoni, Ricerca Sci. 24, 1954, 1039-44

(iii) I.G. Mikhailov and A.A. Chislorazum Doklady Akad Nauk SSSR 81, 1951, 779-82

(iv) P.R. Naidu and V.R. Krishnan J. Phys. Soc. Jap. 20 (10) 1965, 1554-8

88. A.P. Grishin-uchem. Zavedenii 3, 1960, 857-61

89. (i) I.C. Petrec, Sludii Cercetasi Fiz 16(7), 1964, 799

(ii) B.H. Krishnamurthy J.S.I.R. 17 B, 1958, 397.

(iii) Ferninando Demesso, Atti accad. nazl. Lince, Rend classc Sci. Fismal e nat 17, 1954, 370-5

(iv) Feininando Demesso, and Elisa Fadigate ihid 14, 1953, 81-9.

(v) A.V. Kazarjan, L.G. Melkonyan, Premenie ultra akustik R. Issle dovan Veschchestra 9, 1959, 71-81.

90. H.S. Rama Rao and B. Ramchandra Rao J.S.I.R. 20 B, 1961, 157

91. M.G. Seshagiri Rao, and B. Ramchandra Rao J.P.A.P. 3, 1965, 207

92. A. Voleisis, V. Domarkas, E. Jaronis (USSR) Nauch Tr Vyssh Ucheh Zaved, Litov SSR ultrazvult 1969(i), 169-73

93. A.L. Sunder Rao, Bhatnagar, M. Pancholy J.S.I.R. 6, No. 2B, 1947, 35-6

94. P.G. Puranik, Vijaya Kumar, S.N. Murthy Bhagvantham Volume P-13.

95. S.S. Batsanov and A.S. Sonin Krislallografiya 1. 321, 1956.

96. R.J. Fort, W.R. Moore Trans. Faraday Soc. 61 (514), 1965, 2102

97. M.G. Seshagiri Rao, and B. Ramchandra Rao J.P.A.P. 3, 1965, 207

98. (i) Well Harrison, Berg, Ind. Engahem 36, 1944, 871

(ii) Ewell, J. Am. Chem Soc. 63, 1941, 2475

99. I.V. Paracanov. Haudh Doklady Vysshash Kolay Khim-2-Khim Tedkhnol 1959, 240-3

100 S.S. Balsanov Kistallographiga 1, 321, 1956

101 S.V. Subramanyam and K. Ragunath Ind. J. Phys 1969, 278.

102. P.R.K.L. Padmini Trans. Faraday Soc 56, 1960, 1404-8

103. I.G. Mikhailov C.R. (Doklady) URSS 31, 1941, a, 324

ibid ibid 10, 1947, 238

104. G.W. Willard J. Ac. Soc. Am 12, 1941, 438

ihid ihid 19, 1947, 233

105. G.J. Burton J. Ac. Soc. Am. 20, 1948, 186

106. V.F. Nozdrev, N. Larenow Dokl Akad. Nauk S.S.S.R. 92, 1953, 991
107. R. Kuhnikies, W. Schaafs, Akust Beih, 12, 1962, 254.
108. J. Stone and R.E, Pontinen J. Chem. Phys. 47, 1967, 2407.
109. A. Glacomini. J. Ac. Soc. Am. 19, 1947, 701-2
110. Srinibas Pande, Bihuti Bhusen Ind. J. Phys. 31, No. 4, 1957, 227.
111. R. Kuhnikies, Dissertation an dor TV Berlin 1962
112. R.N. Barfield, W.G. Schneider J. Chem. Phys. NY 31, 1959, 488
113. B. Jacobson. Ark Kemi 2, 1950, 177
114. R. Kuhnkies and W. Schaafs Acutica 12, 1962, 254
115. ibid ibid 13, 1963, 407
116. G.W. Marks J. Acoust. Soc. Am. 27. 1955, 680
117. I.G. Michailow. D&N 31, 1941, 324, 550
118. P, Prosorow. Zh. Fiz. Khim 14, 1940, 391
119. C.R. Randall Bur Std. Rep. 402, 1932, 8, 79
120. J.C. Mc. Dade, D.R. Pardue, A.L. Hedrich and F. Vratarie J. Acous t. Soc. Am. 31, 1959, 1387
121. I.G. Mikhailov. Proceedings of fourth international congress in Acoustics J. 43, 1962
122. W. Schaaffs Acustica 6, 1956, 382
123. N.I. Koshikin, Primen Ultraakust Issled Veschchesta No. 22, 1967 29-32
124. R.T. Lagemann, L. Gilley, E. Mc Perof, J. Chem. Phys. N.Y. 21, 1953, 819
125. A. Borone, G. Pisent and D. Sette Acustica 7, 1957, 109
126. S. Smurnov, Dissertat siya (Thesis), Moscow State University, 1952.
127. A. Chynowoth and W. Sehneder J. Chem. Phys. NY 19, 1951 1566, 29, 1952, 1777
128. M. Cevolani and S. Petralia Nuovo Cemento 1, 1955, 705.
129. J. Frenkel, Kinetic theory of liquids Oxford Univ. Press Ltd. London, 1946.
130. Alfred Weissler J.A.C.S. 71, 1949, 1272.
131. Theory of rate processes by Glasstone, Laider Erying Published by Mc. Gzaw Hill New York 1951 P (—) 488
132. Telang J. Chem. Phys. 17, 1949, 556
133. S.R. Palit J. Chem. Soc. Part UUU, 1956, 2740
134. D.B. Macleod Chemical Abstract 19, 2151
135. Ashok Kumar Mukerjee, J. Ind. Chem, Soc. 28, 1951, 363.

Index